MW01596836

# Alternative Careers in Sci-Tech Information Service

The *Science & Technology Libraries* series, Ellis Mount, Editor:

Vol. 7

*1 Innovations in Planning Facilities for Sci-Tech Libraries*
*2 Sci-Tech Libraries Serving Societies and Institutes*
*3 Preservation and Conservation of Sci-Tech Materials*
*4 Alternative Careers in Sci-Tech Information Service*

Vol. 6

*1/2 Sci-Tech Libraries in Museums and Aquariums*
*3 Weeding of Collections in Sci-Tech Libraries*
*4 Role of Computers in Sci-Tech Libraries*

Vol. 5

*1 Serving End-Users in Sci-Tech Libraries*
*2 Fee-Based Services in Sci-Tech Libraries*
*3 Role of Maps in Sci-Tech Libraries*
*4 Data Manipulation in Sci-Tech Libraries*

Vol. 4

*1 Role of Serials in Sci-Tech Libraries*
*2 Collection Development in Sci-Tech Libraries*
*3/4 Management of Sci-Tech Libraries*

Vol. 3

*1 Online Versus Manual Searching in Sci-Tech Libraries*
*2 Role of Translations in Sci-Tech Libraries*
*3 Monographs in Sci-Tech Libraries*
*4 Planning Facilities for Sci-Tech Libraries*

Vol. 2

*1 Current Awareness in Sci-Tech Libraries*
*2 Role of Patents in Sci-Tech Libraries*
*3 Cataloging and Indexing for Sci-Tech Libraries*
*4 Document Delivery for Sci-Tech Libraries*

Vol. 1

*1 Planning for Online Search Services in Sci-Tech Libraries*
*2 Networking in Sci-Tech Libraries and Information Centers*
*3 Training of Sci-Tech Librarians and Library Users*
*4 Role of Technical Reports in Sci-Tech Libraries*

# Alternative Careers in Sci-Tech Information Service

Ellis Mount
Editor

The Haworth Press
New York • London

*Alternative Careers in Sci-Tech Information Service* has also been published as *Science & Technology Libraries*, Volume 7, Number 4, Summer 1987.



The Haworth Press, Inc., 12 West 32 Street, New York, NY 10001
EUROSPAN/Haworth, 3 Henrietta Street, London WC2E 8LU England

**Library of Congress Cataloging-in-Publication Data**

Alternatives careers in sci-tech information service/Ellis Mount, editor.
p. cm.
"Has also been published as Science and technology libraries, vol. 7, no. 4, summer 1987" – T.p. verso.
Includes bibliographical references and index.
ISBN 0-86656-694-5
1. Librarians – United States – Employment. 2. Information scientists – United States – Employment. 3. Library science – Vocational guidance – United States. 4. Information science – Vocational guidance – United States. 5. Science – Information services – Vocational guidance – United States. 6. Technology – Information services – Vocational guidance – United States. 7. Career changes – United States. I. Mount, Ellis.
Z682.2.U5A44 1987 87-22317
020'.23 – dc19 CIP

# Alternative Careers in Sci-Tech Information Service

Science & Technology Libraries
Volume 7, Number 4

## CONTENTS

# Alternative Careers in Sci-Tech Information Service

# Introduction

Many librarians working in the areas of science and technology have probably given little thought to careers involving handling sci-tech information that lie outside the traditional realm of librarianship. Less surprising, perhaps, is the fact that many scientists and engineers have also not been very much aware of alternative careers for people like themselves. Yet there are several professional positions which involve science and technology that fall outside traditional career fields.

The purpose of this issue is to point out to those having a sci-tech background, whether they be librarians or practicing scientists and engineers, something about alternative careers. Some of the positions described here may actually demand more of people than do many traditional jobs, but the rewards may also be greater.

It is hoped this issue will help fill the gap in the literature on careers in alternative fields for those wanting to work in the disciplines of science and engineering.

Seven alternative careers are described, beginning with information broker, written by Minnie L. Johnson, who manages her own company in this field. The next paper, by Astrid Johanson, is on translating; she is a senior technical translator at AT&T-Bell Laboratories. The work of an acquisitions editor is described in the next paper by James L. Smith, who is an acquisitions editor at John Wiley & Sons, Inc. The fourth article discusses the information resources manager, the position held at Johns Hopkins University's Applied Physics Laboratory by the paper's author, Wilda B. Newman.

The following paper covers the career of a researcher in sci-tech information, written by Dr. Robert E. Stobaugh, Manager of the Research Department in the Information Systems Organization at Chemical Abstracts Service. The career of managing an online database is described by Taissa T. Kusma, who performs that work for the American Mathematical Society's database known as Math/Sci.

The seventh paper deals with careers as abstractors and indexers, written by two employees of the American Petroleum Institute's

 

Central Abstracting and Indexing Service, namely Ellen Young (Senior Abstract Editor) and Elliott Linder (Senior Editor for Technical Indexes).

There are two special papers for this issue. The first describes an outstanding collection on electricity which was given to the Engineering Societies Library by the Institute of Electrical and Electronics Engineers; the nature and preservation of the collection are discussed. The paper was prepared by Ronald R. Kline, Joyce E. Bedi and Thomas D. Lindblom, writing for the IEEE Center for the History of Electrical Engineering.

The other special paper continues the preservation theme of the previous issue (vol. 7, no. 3), being an account written by John P. Baker, Assistant Director for Preservation at the New York Public Library; he covers preservation of the collection of the Science and Technology Center of that library.

The collections study for this issue deals with the literature and history of Computer-Aided Design/Computer-Aided Manufacturing (CAD/CAM), written by Colette O'Connell, Engineering Librarian and Coordinator of On-Line Search Services at Rensselaer Polytechnic Institute. Part II of this paper will be found in the next issue of this journal (vol. 8, no. 1).

Our regular features complete the issue.

*Ellis Mount*
*Editor*

# Information Broker: A Career in Scientific and Technical Information Service

Minnie L. Johnson

**SUMMARY.** The role of an information broker is comparable to any other broker, i.e., real estate, stock, commodities, in a competitive environment. The only difference is that the product is information. Information is researched, purchased, packaged and distributed to a select clientele for a fee. The role of the information broker is challenging and nontraditional because it demands a concentration on the generalities of the business arena as applied to the stringent data collection requirements of the research scientist.

## *INTRODUCTION*

Where do I go from here? This is the dilemma many assertive library and information professionals ask after a few years in the field. At each conference or meeting, many professionals pose the question of "how can I change the image of the Librarian, gain respect from administration and be given the complete responsibility of a manager or as a 'shaker and mover' in the industry?" Professionals in the sci-tech environment must think of themselves as researchers. A researcher looks for a void, analyses the void, outlines steps for filling the void, proceeds to apply solutions, tests the solutions, examines the results, formulates and documents the conclusions. Finally, the researcher presents the results and conclusions to associated colleagues in the environment. The professional must become research oriented as an information specialist, embrace the team concept to perpetuate the growth of the product, the company and self. Buccherl describes the function of an information

Minnie L. Johnson is Information Broker. Her company is SCI-Tech Information Services, 12 Bank St., Summit, NJ 07901. She received the PhD and MLS degrees from Rutgers University, and an MA (Physiology) from Hunter College.

 

specialist as a change agent, because they must recognize performance deficits in the environment and deal with the deficiencies. They must attempt to influence innovative decisions in the direction that the corporation and/or the . . . profession feels is desirable.[1]

## SKILLS

### Academic Skills—Hard Sciences vs Soft Sciences

Any subject which man can study by using the scientific method, an orderly system of solving problems, may be called a science. The sciences include (a) mathematics and logic; (b) the physical sciences; (c) the biological sciences; (d) the earth sciences; and (e) the social sciences. The hard sciences are divided into the pure and applied sciences. Pure science summarizes and explains the facts and principles discovered about the universe and its inhabitants. Applied science uses the scientific facts and principles to make things that are useful to man. The soft sciences (social sciences) are so defined because they are concerned with the study of man and society and are difficult areas to investigate.[2] It is hard to identify laws or principles when dealing with human behavior.

It is imperative in a scientific environment for the professional to have an undergraduate degree in a hard science. This helps to ensure that the professional thinks, reacts and operates as a research scientist in an information and library arena. The professional can communicate with scientists and understand their needs for information. It has been my experience that the time to train the hard science candidate in information retrieval is considerably less than the time it takes to train a social scientist to acclimate to the technical and hard science environment requirements. Although both can operate and obtain information, the time in interpreting the query, analyzing the data and presenting the results in a format that is acceptable to the client depends upon the subject specialization of the performer.

### Professional Skills

Experience is always the best teacher. There are permanent analytical skills that are acquired while working in a laboratory and on research projects that are applicable to the sci-tech information environment. These skills involve formulating the experimental design, objectives and goals, and projecting the results to be expected. Dalton[3] describes the skills necessary for a subject specialist as:

— to collate dispersed information
— act as a catalyst for the exchange of information
— communicate new developments
— gather and provide informed comments on these developments
— establish files of facts and events
— provide up to date "state of the art package" summaries at short notice
— analyze information and detect emerging patterns

Therefore, research exposure is significant when one is communicating with research scientists and physicians. It puts the client at ease because the operational definitions are understood and are compatible. The technical association makes the pre-search interview less traumatic. It also indicates that the professional is interested in the pursuit of knowledge as well as directing the flow of knowledge. The professional practices the discipline of being a part of the research team to the completion of the experiment, in addition to showing where the information can be found.

Some of the project areas that are available where a professional can participate as a member of a task force are described by Faust[4] and expanded here based on personal experiences.

### *New Chemical Entities*

A patent is the first literature documentation for a product in research. It discourages the re-invention of the wheel and is most cost effective. The monitoring of a NCE from patent to market and post-market surveillance can create an invaluable communications network for the professional, chemists, biologists and pharmacologists. The needs of these scientists are for new research trends, historical and current R&D data.

### *Safety*

Environmental scientists need information on the permissible exposure limits of various chemicals. Pathfinders and on-demand information are constantly required for preventive exposure and acute exposure.

### *Drug Regulatory Affairs*

Quarterly and annual documentation updates are required for FDA submission on all consumer products. This information is usually part of the public domain and can be easily monitored via an online system. The clientele consists of both physicians and scientists.

### *Product Management*

This area is always looking for a better way to package a product. Competitive product intelligence is an ongoing process and can provide the competitive edge for a company when delivered in a timely fashion. Requests for information are from administrators, writers, clinicians, sales representatives and business personnel, because their needs are for knowledge of leads in other firms, academic institutions and government research organizations.

### *Adverse Drug Reactions*

Close monitoring is necessary for reporting drug reactions and deaths to the FDA. Lots of time in the reporting of ADRs can result in a penalty to the company and a temporary loss of credibility. A relatively close liaison has to be established with clinicians.

## *ENVIRONMENTS*

### *Protected Environments*

The protected environment is one where the professional is in a corporate or quasi-corporate environment.[5] Although, the professional is accountable to the client, a mistake or error can be rectified without loss of personal credibility. A case in point is information required on product intelligence. A search is performed, analyzed and submitted to a client. After a review by the client, the search may have to be repeated. The time is lost and the cost duplicated, but this time and cost loss is absorbed by the entire unit. The client may still be dissatisfied with the results, but will continue to use the resources of the professional because the information is free. However, the position, credibility, and accountability of the professional is still intact.

### Unprotected Environments

In the unprotected environment, the professional operates as an independent and is in business to provide information for a fee.[6] Any misinformation or error is absorbed by the Information Scientist/Broker to the point where one may consider the statement "if not satisfied with the results, the information and services are rendered free." This statement gives accountability to the professional services, but cuts into the profit margin if too many types of these performances occur. Plus, the position and the credibility of the Broker is threatened because of the competitive business environment. This professional is required by necessity to provide accurate fee-based information support and be accountable to the client in order to obtain repeat business and referrals.

## OPERATING THE BUSINESS

### Formation of a Business

Some of the best literature on starting a business is supplied by the Small Business Administration.[7] One of the first analyses to perform is to ascertain if you are serious and ready to go out into the "real world," or if are you still in the planning stages. The next big step is to develop a business plan. The plan will assist you in identifying your needs and making many concrete business decisions such as:

1. Projected goals
2. The format of the business (sole ownership, partnership, corporation)
3. Product(s) or service(s)
4. Homebased or office
5. Type of equipment, if necessary
6. Demographics (geography, clientele)
7. Financial planning (start-up costs, loans, professional services and personal needs assessment)
8. Related activities (administration, marketing)

The business plan can be tested on a small target before you go solo because a cash flow problem can develop almost immediately unless you have unlimited resources or a substantial client base.

Even scientists must be educated to obtain information outside of their normal spheres. The plan also serves as a reminder of the goals set for the business. Therefore, parts of the plan may have to be modified every 3-6 months, especially as to the products, services, geography and clients.

### *Maintenance of a Business*

Once the business is established, the professional should begin to concentrate on the maintenance of the business as a commitment. The professional, at this stage of the process, must become more assertive, and obtain success by becoming an independent Information Scientist/Broker. The scientific research skills and the newly acquired business orientation must enhance each other. The main focus is now on the ability to sell the skills, products and services to the client base. Thus, the professional has to operate in a combination of arenas, as a scientist, and as a business entrepreneur who is required to intensify efforts in the following aspects of the business:

1. Identify certain scientific areas and be knowledgeable about the areas where you can make an impact—chemistry, medicine, engineering, etc.
2. Know your product and/or services.[8] If you are unsure of what you are offering, it is exceedingly more difficult to convince your clientele.
3. Be aware that time management and the 'follow through' concept are essential. Scheduling time slots and making effective use of that time make an important difference. 'Follow through' appears to be germane only to sports activities, however it is a basic concept in hitting the target in any activity, especially in business.
4. Join and become actively involved in related associations by networking and exchanging information and your services.[9] Many associations provide minicourses in business survival, especially on what you thought you should have known. Consult with experienced business associates and do not be embarrassed to ask for help in areas where you are not well versed.
5. Maintain a budget for continuous advertising and marketing because the best product will not sell if no one knows that it is available.

6. Delete 'failure' from your vocabulary for each experience can be thought of as a learning vehicle by which you can avoid other pitfalls.

## *SUMMARY*

Magic strategies are not available for making a smooth and positive career change, and the literature cannot make you aware of all the minute details that need attention. The true test of the professional acumen of the Information Scientist/Broker is when you approach each task as if your life depended on it. When you find that it actually does, remember that life is a continuing process of education where one has to set goals, work hard, actualize potential, and aspire.

## REFERENCES

1. Buccherl, Paul; Baker, John A. Management strategy for the diffusion innovation: unit dose distribution. *American Journal of Hospital Pharmacy*, 35(2):168-173, 1978 Feb.

2. McGrath, William E. Relationships between hard/soft, pure/applied, and life/non-life disciplines and subject book use in a university library. *Information and Processing Management*, 14(1): 17-28, 1978.

3. Dalton, Micheal S. The role of subject specialism in the future development of information services. *Journal of Information Science*, 1(2): 107-112, 1979 May.

4. Faust, R. E. Information needs at the research/corporate management interface. *Drug Information Journal*, 10(1): 13-15, 1976 Jan.-Mar.

5. Jackson, Eugene B., ed. *Special librarianship: a new reader*. Metuchen, NJ: Scarecrow Press, 1980, chapter 3.

6. Maranjian, Lorig; Boss, Richard W. *Fee-based information services: a study of a growing industry*. New York: R. R. Bowker, 1980, chapter 8.

7. U.S. Small Business Association. *Management assistance publications*. Washington, DC: SBA, 1985.

8. Sellen, Betty-Carol; Berkner, Dimity S. *New options for librarians: finding a job in a related field*. New York: Neal-Schuman, 1984, part 2.

9. Warnkern, Kelly. *The information brokers: how to start and operate your own fee-based service*. New York: R. R. Bowker, 1982, chapter 8.

# Translator: A Career in Scientific and Technical Information Service

Astrid Johanson

**SUMMARY.** Requirements for success as a technical translator are discussed. The role of the computer is analyzed, as is the outlook for machine translation.

## *SKILLS AND TALENTS*

Good technical translators, the professionals, are made in heaven. Nevertheless, certain very necessary skills, which a professional translator must have, are added here on earth. Expertise in languages alone is not adequate. There is a very special mixture, consisting of talent for languages, good writing and communication skills in two or more languages, an innate curiosity, liking and thorough understanding of scientific disciplines, which makes a professional technical translator. All of these talents are indispensable and can make or break the career of a professional translator.

Languages are not "picked up" in a few easy lessons, and an education in the sciences as well requires even more study, yet both areas need to be studied in great detail for a career in technical translating. These studies, by necessity, are lengthy and curriculum emphasis is usually placed on language studies for translating. Curricula, which include the teaching and training in translation came into being about a decade ago. Before that, in-house translators in large corporations were trained on the job, usually by a senior in-house translator, if such a person was present.

I started my career as a technical translator at AT&T Bell

Astrid Johanson is the Senior Technical Translator, AT&T Bell Laboratories, Murray Hill, NJ 07974.

 

Laboratories several decades ago. I have the appropriate ''mixture'' and prior experience in translating. I was also very fortunate to have been trained by the Senior Technical Translator in residence. By any standards, she was a very strict taskmaster indeed; however, now, years later, I am grateful for the fine art of translating, technical writing, and communication this extraordinary person taught me.

## *IMPACT OF COMPUTERS*

It is quite clear, that the computer, either a PC or a terminal, has had far-reaching effects on all of us and has significantly changed our lives, our everyday language, even the way we think and perform some tasks. The translating profession, too, has not only been impacted, but has undergone certain changes, by the computer. And, certainly, many new doors have been opened to translators.

The only thing that should be said about the ''old days'' is that before the advent of the computer almost all of the translating tasks (except interpreting) were carried out manually. The only ''automated'' exceptions were the dictaphone, which was, and still is, favored by many translators, and the electrical typewriter. Other than that, everything was a manual process—tedious, time-consuming, and, at times, repetitive and even boring.

As I mentioned above, the computer has quite significantly impacted the translating profession. Even in their ''infancy,'' PCs and terminals offered the advantage of word processing software. Word processing did away with tedious retyping and text correction. It allowed various types of formatting and easy handling of formulas, equations, and tabular material. PCs and terminals continue to become more and more sophisticated, offering more sophisticated software with a multitude of very useful options. The latter can be easily adapted to translating tasks, increase the translator's output, offer various attractive formatting packages and communications.

With the appropriate software, the translator can create and store online his or her own dictionaries of specialized terminology in several languages, eliminating time-consuming reconnaissance of hard-cover dictionaries and the collection of terms which cannot be found in dictionaries. This certainly is superior to collecting 3 × 5 cards in shoeboxes, a favorite occupation of translators. Online dictionaries are easily searched and updated; obsolete terms are easily deleted. Look-up is easy while working on a translation and can be carried out in several ways, depending on the software used. In a computer assisted translating environment, the translation of a document is directly produced online, i.e., directly from the source text by the translator. Errors are corrected either immediately or later; there are no limits to editing—text can be added, changed, paragraphs or sentences moved from one location to the other and so on. Such changes can be made any time, as long as the translation of a document is in a file, stored in the computer memory. Tedium is removed, working speed is increased, and many new options beckon the translator. Learning a word processing system is very easy and should not take too much time.

Generally speaking, this process is known as Computer Aided Translation (CAT). Computer aids for translators are being developed and marketed at an increasing rate these days. In using a CAT system, the translator is in complete command of the translating process, and the produced translation is a product of the translator's intellect and expertise as well as of the options available in the software. CAT has become quite popular with the translating profession. Many of us translators have also learned a few basics of programming and have customized our software to meet our own specific needs.

Another, sophisticated option offered by the computer is machine translation (MT). MT has been around for more than two decades, but in its early stages it was not very promising or efficient. There have been significant changes in the MT field—hardware as well as software have become rather sophisticated and the linguistic approach to natural language translation has changed.

MT has had a significant impact on the translating profession. In my opinion, it has broadened our professional field by not only extending our capabilities to handle very large amounts of work within a short period of time, but has also added new avenues for translators.

It is quite clear that no matter how good or sophisticated an MT system might be, it will not work without the involvement of human intellect. On the one hand, the necessary hardware and software to handle the translation of one language to another must be developed. On the other hand, the human translator's intellect and expertise must be involved to provide the necessary dictionaries residing in the software, which are indispensable for translation output (the machine does the dictionary look-ups) and post-editing.

At the present time, the accuracy of MT output is claimed to be about 80% to 85%, which means that post-editing of raw MT output is required. So, translator to the rescue! Post-editing raw MT output is an additional new skill for translators and is very rewarding indeed. The machine has performed the "rough spade-work"; the translator provides the final touch, the necessary accuracy, error correction, and the elegant turn of the phrase.

MT, therefore has opened two new avenues for translators. These avenues are highly professional—that of the terminologist and that of the post-editor.

Of the many arguments against MT, some are valid, some not. There are certain types of documents which do not lend themselves to being processed by MT, such as patents, journal papers and so on. But there are many documents which can be translated with the aid of MT—such as equipment manuals and training manuals. The important factor in such cases is the speed in which this type of translation can be handled with MT, even into several languages simultaneously, if need be. As a rule, this type of documentation is needed for marketing and sales ventures, where translations must be available within very short periods of time and manual processing of translations is too time-consuming and not cost-effective.

There is a type of translation, however, which can be handled on MT, although the document itself does not easily lend itself to this type of treatment. This is the "need to know only" translation, i.e., instances where only the gist of the document is needed and a certain amount of accuracy and correct grammar can be sacrificed. In these circumstances, MT is a tremendous time-saver and a cost-effective way of determining the content of a document.

## *FUTURE OUTLOOK*

Having briefly mentioned the past and having described the present in some detail, a look into the future is in order. Building on the present, I can envision a very interesting future for the translating profession. Computer aids for translators will become increasingly sophisticated and, at the same time, more user-friendly. Software supporting translation work is being developed already; such developments will continue. Interest in MT has been renewed; it will prove to be a very useful as well as cost-effective tool in the future.

The importance or involvement of the human translator will not diminish. Human translators have been actively involved in the transfer of information as well as cultures since time immemorial. Now, that we are living in the information age, their role is an even more important one indeed. Not only are translators involved in information transfer itself, but since the computer has relieved them of many tedious and mechanical tasks, they can assume an active role in database searching, information retrieval, post-editing, and online dictionary construction. Their translating expertise will allow them to access databases which are not in the English language, prepare a translation of an abstract online, and transmit this information to their customer via electronic mail.

Another important and active involvement for translators is in the area of international business protocol. Large corporations are very much involved in the global marketplace. Translators will become increasingly involved in disseminating information in such areas as the cultural background and customs of certain countries, the protocol of conducting business negotiations in these countries, and so on. Translators, because of their multi-lingual and often multi-national background or education can make important contributions by supplying information about the very many factors involved in successful, global business negotiations. For example, the translator prepares country briefing packages containing pertinent information about a specific country where the corporation wishes to develop business and marketing interests. These packages can be supplemented with foreign company profiles and various other information in order to meet special marketing needs. The translator also prepares newsletters of foreign or domestic competitor information, disseminated from online databases, the domestic or the foreign press, and other sources. The translator's language expertise is an asset in this case. Market and competitor oriented newsletters can cover a range of products, certain companies or are targeted on a certain country and market, depending on the corporation's marketing interests.

Large corporations and Government agencies employ protocol officers, whose function is to provide the proper protocol environment for foreign business negotiations held in the United States as well as abroad. The translator and the interpreter help to overcome the language and the cultural barriers, working in close contact with the protocol officer.

It is clear, that the translator is no longer someone hiding behind a large stack of dictionaries, waiting for documents to be submitted for translation, and tediously cranking out translations of documents. The translator of the present and of the future is a person liberated by the computer and is very much involved in information transfer in its very many different facets, someone who builds bridges not only between different languages, but between the many different cultures as well, and actively participates in determining organizational strategy.

I thank my colleagues in the Central Information Services group of the AT&T Library Network Operated by Bell Laboratories who have supported my many linguistic activities. I also thank Donald T. Hawkins, who gave me the idea of combining my translating expertise with database searching.

# Acquisitions Editor: A Career in Scientific and Technical Information Service

James L. Smith

**SUMMARY.** A discussion is given of the Acquisition or Sponsoring Editor's role in the handling of contemporary scientific and technical information, including the background and training that are desirable, whether it be subject matter expertise or a publishing background. The contemporary situation is described relative to scientific and technical data, print, monographs, series volumes, encyclopedias and electronic disks, and databases. Other topics include: the creative writer of fiction and the scientist/technologist as writer, the computer revolution evaluated in traditional print publishing, the importance and desirability of the profession of publishing, the varied skills that are utilized in the handling of scientific and technical information today.

## *BACKGROUND AND EDUCATION*

Traditionally, the "Acquisitions" or "Sponsoring" Editor for advanced reference materials comes to his/her job via one of three routes. This person either has subject matter training in the discipline being handled or publishing experience or a combination of both. The focus of this article will be on the Acquisitions/Sponsoring Editor's role in the contemporary science/technology information situation.

For the type of reference level, advanced, state-of-the-art sci/tech

---

James L. Smith is Acquisitions Editor in the Scientific/Technical division of John Wiley & Sons, Inc., 605 Third Ave., New York, NY 10158. His twenty years of experience in Sci/Tech publishing have included college textbook work in mathematics, physics and engineering, in addition to his primary involvement with chemistry and chemically related subjects over the years. He earned his undergraduate degree at Boston College and did graduate work in English and American Literature at the University of Wisconsin.

publishing to be dealt with here (since this is presumed to be of interest to readers of *Science & Technology Libraries*), it would seem that subject matter expertise is indeed a prerequisite. This may still be true, but my own experience as well as my view of the changing mix of systems in the processing of sci/tech data have convinced me that formal academic training may be only part of the story.

For print, deciding what is to be published remains primarily a quantitative decision. Monographs, series volumes, encyclopedias or handbooks are planned and executed by the Acquisitions Sponsoring Editor generally after discussions which may involve consulting editors or advisors or subject matter specialists on the outside and marketing/design/production people on the inside. Even if the idea for the book begins in the mind of the Acquisitions Editor, because this individual "knows" the subject (the more specialized the monograph, the less likely this will be), nevertheless, the prudent Editor would not consider committing to the project without reviews and/or consultations with his advisor, etc., to say nothing of his or her own marketing people and publisher.

The four types of books mentioned above: monographs, series volumes, encyclopedias and handbooks, all are involved in this quantitative decision making process. However, the degree of subject matter expertise brought to bear on the part of in-house editors varies depending on the nature of the print being handled. A monograph or contributed volume in a series may well be launched virtually on the consulting editor's or advisor's decision alone. It would be difficult to imagine, however, a multi-volumed encyclopedia or large handbook in any science or technology being successful without having hands-on subject matter editing within the publishing house. This presupposes subject matter expertise in the discipline in question.

## *MEDIA OPTIONS*

Work in the Sci/Tech information/communications area today and presumably in the future involves more than print. The subject matter decision making process becomes less clear when one attempts to deal with something other than print. Positive or negative reviews of partial or complete manuscripts are one thing. The delivery system (print) has already been decided. Additional decisions are necessary when the delivery system can most efficiently be handled in any one of several electronically retrievable forms: hard disks, floppy disks (either standing alone or in tandem with print), or databases. This consideration gives a new dimension to the original question of subject matter expertise and the role of the Acquisitions/Sponsoring Editor.

As indicated earlier, the Acquisitions or Sponsoring Editor may have an advanced degree in the discipline that he/she is handling or this person may have gained experience in the publishing business by way of serving as a traveler, representative or sales person and later Acquisitions Editor in the College Division of a Sci/Tech publishing house. It has always seemed to me that the question of a preferred background for a successful Acquisitions Editor has been moot. Distinctions are even more blurred today given what has been called the "Electronic Challenge" to traditional print modes.

Sci/Tech publishers now utilize the skills of computer people in planning the most efficient delivery system for the information. These "Electronic Publishing" people may be, in one sense, subject matter specialists in that they know computerized information processing, but decisions as to what material is to be handled electronically, packaged and marketed still involves the editorial-editing production team function of traditional print.

There are many skills involved in the handling of Sci/Tech information today whether print or electronic. Editing people, editors, translators, indexers, database managers, production people, marketing people and librarians all share involvement in one of the most dynamic aspects of contemporary human activity: the dissemination of scientific and technical information. The exponential growth of this information demands more varied skills on the part of all these people, but the stimulation derived from involvement with this work is, in my opinion, without equal.

These many and varied skills certainly are brought to bear in the publishing industry. The changing technology of traditional book publishing (computer composition, new binding methods, etc.) parallels in a less dramatic fashion, the tandem technologies of database information retrieval or electronic publishing. The publishers have hundreds of years of experience in synthesizing, organizing and packaging information, and the publishers should be the logical leaders in the exploitation of new methods for synthesizing, organizing and packaging the delivery systems for this information.

## SKILLS

The necessity of utilizing these additional skills in today's multimedia publishing industry would indicate that there are definite advantages to having a sci/tech background for this type of publishing. Someone with a sci/tech background would find it beneficial (by virtue of his/her training) in that he or she brings some of these skills to bear. A scientifically trained individual would find that in the initial stages of his/her publishing career an advanced degree would be helpful. However, in the long run, given the complexities of the publishing industry today, the general, non-formal skills gained through experience in publishing will be of equal if not more importance to this person's success as an Acquisitions/Sponsoring Editor.

I have alluded before to the dynamism that I find in sci/tech publishing. Personally, I have had little formal scientific training, but long ago a conscious career choice was made on two compelling ideas: a life-long love of books and a desire to be part of the excitement of advances in science. In one sense, this may have been a reaction to my formal education in the liberal arts, but personally, I have always found science and technology at the center of my worldview.

## ROLE OF THE ACQUISITIONS EDITOR

A great deal of time in my literary pursuits has been devoted to attempting to understand how the creative writer of fiction can fashion a "new" reality, and in so doing, if he be successful, evoke a universal response. The physical laws, the physical universe, the so-called space-time continuum, can be violated with impunity. Reality may be distorted or minimized, warped, expanded, broadened, narrowed or telescoped. Reality can become surreal. The ways that the universe may be manipulated are inexhaustible as long as the writer is talented.

The creative scientific or technological investigator on the other hand, seems to me to approach reality in a linear fashion. He or she adds to, refines, or revises our knowledge of the perceivable universe and unlike the literary investigator, achieves universality if the facts are "true" and thereby validates reality. Just as the fiction writer's universe is inexhaustible, however, the scientific investigator's quest for answers to questions lead to still more questions and answers in an endless chain.

The Acquisitions/Sponsoring Editor certainly is an intimate part of this process. The excitement felt by the Author/Editor of a monograph or contributed volume when the book appears in print is shared albeit in a less intense way, perhaps, by the in-house Editor and Editing/Production people. Post-publication reviews are read avidly and incorporated into the book's advertising and promotion where appropriate. It is sometimes possible to track sales which are directly linked to positive reviews in journals where the book is advertised by way of reader service cards. It is still a great pleasure and gets to the heart of the matter, for the Acquisitions/Sponsoring Editor to read positive reviews of his/her books.

As indicated earlier, the feeling in the industry currently is that non-book products and services such as educational software and database publishing will have an impact on traditional book publishing as it relates to Sci/Tech disciplines. This could be viewed negatively as a threat to traditional book publishing. But in my opinion, the more astute management of traditional publishing companies will react in a more positive way. Strategic planning will have to be more closely aligned relative to the direct needs of the market in terms of other possible delivery systems that may do the job of getting the information across better than monographs, encyclopedias, or handbooks.

The danger, of course, is overreaction. There have been recent examples of premature investment on the part of publishers in systems that have failed badly. My own view is that the computerization will impact on traditional publishing but the changes that come about will be evolutionary rather than revolutionary. I sense that there is a great deal of information available to me to be retrieved electronically and conceivably this will eventually alter the way I function in my job of determining what my company publishes in the disciplines for which I am responsible. For the near term, however, the rapid development of better hardware and particularly more functional software in the personal computing field is so fast-paced as to outstrip the bracketing of that field.

Much of the quantitative data of traditional print publishing in the Sci/Tech area lends itself to the notion of database creation and this activity is indeed proceeding apace. In some cases, this is successful, i.e., there are customers who subscribe to the database because they can make economical and efficient use of retrieving the information electronically. On the other hand, in a few cases, we are actually seeing the reverse of this. Successful books are being printed from data originally stored electronically.

The example I have in mind is toxicology. The last twenty years or so have witnessed the extensive proliferation of databases for toxicological data both in the government and in the private sector. In some cases, some of this information has appeared in suitably altered book form thereby completing the circle. The danger for contemporary investigators in this and other areas is to slide into the trap of believing that if the information is not computerized, it doesn't exist. The quantification of toxicological data was going on before we had computers just as printing existed before the typewriter.

These systems complement one another. Almost twenty years ago, I listened to a dinner speaker talk about computers and publishing. This man was an electrical engineer involved with computers and he was at pains to describe what he considered to be the ideal information storage and retrieval system. Such a system would contain X number of bytes of information, it would be (relatively speaking) inexpensive, it would be portable and (again, relatively speaking) indestructible. This ideal system was, of course, the book.

A great deal has changed in the last twenty years and much will change in the next twenty. The paperless office hasn't happened yet, at least not in my experience. In fact, the opposite seems closer to the mark. There are stacks of computer printouts everywhere, and our dependence on computers or better still our use of computers will become even more extensive. It behooves the publishing industry and those who are part of it to rise to the challenge. I believe that the next twenty years will be more like than different then the last twenty and the handling of scientific and technical information will continue to delight and challenge those of us who are part of it.

# Information Resources Manager: A Career in Scientific and Technical Information Service

Wilda B. Newman

**SUMMARY.** In recent years new careers have come about as a result of the information age and an increasing demand for a greater variety of information workers. Changes in the demand for knowledge workers coupled with an increasing desire of librarians to become more involved in information-related professions is changing the role of librarians. There is, however, a need for library and information science schools to prepare the new knowledge workers for this century and beyond.

## *CAREER OVERVIEW*

The Johns Hopkins University, Applied Physics Laboratory has been my employer for some 20 years. The work here is primarily R&D as well as test and evaluation in the fields of physical and engineering sciences, particularly physics, mathematics, and geophysics; the computing and environmental sciences; and aeronautical, mechanical, electrical, and biomedical engineering.

The first 16 years of my career included a support position in the translations function, interlibrary loans and reference, supervisor of acquisitions, and supervisor of technical services in our central library. Besides the obvious library activities involved, each position included management at various levels of the library, such as personnel evaluations, budgeting, and computer-related or computer-dependent operations.

---

Wilda B. Newman is Information Resources Manager for the Administrative Services Department, Johns Hopkins University, Applied Physics Laboratory, Johns Hopkins Road, Laurel, MD 20707. Ms. Newman has a BS in Business Management, University of Maryland, and a Master's Degree in Library and Information Science, Catholic University of America. Prior to her current position, she worked for The R. E. Gibson Library, at the same institution, in interlibrary loans and reference, supervisor, library acquisitions, and supervisor, technical services.

 

For the last five years I have been with the Administrative Services Department. First, as Staff Assistant to the department supervisor, evaluating organizational publications and coordinating and maintaining the updating of the Practices and Procedures Manual. Later, as Assistant for Records Management and Procedures, I became more involved in the issues and problems relating specifically to the department and its role in the parent organization. Both assignments contributed to my knowledge and expertise in my current position as the Information Resources Manager for the department.

## *BRIDGING OLD AND NEW CAREERS*

In each of these jobs, I brought with me certain knowledge, skills, and expertise. My 16 years as a special librarian in the organization provided me with a unique background to take on my new assignments. Cortez[1] refers to special librarians as the entrepreneurs of the library profession. He draws attention to the Special Library Association's (SLA) motto "Putting Knowledge to Work," as its particular philosophy and sees this at the root of what today is called Information Resources Management.

On the other hand there was a need for me to become even more aware of the organization, its policies, procedures, and behavior, from the perspective of the Administrative Services Department in the context of the larger organization.

One of the most important steps to take in bridging your old career with a new one is to find out as quickly as possible where the gaps may lay in your knowledge and what expertise you can use immediately to help in finding solutions for your new organization. You can do this through formal and informal interviews, reviewing organizational documents, and discussing specific functions with supervisory staff and others in the organization.

We all become somewhat afflicted with tunnel vision, brought about by working in the same "restricted" environment over a period of time. It is a rare person indeed who knows what it is like for the person on the outside of things, without having "walked in their shoes." Switching sides, so to speak, forces us to make new considerations. Role playing is an example of how managers today are helped to see more of the whole picture than perhaps is normally possible for them to do.

It is important to demonstrate the capacity for seeing the big picture, not simply the local or closest view; to be creative in approaching problems and their solutions; to communicate your findings through good written and verbal communication; and to do these things with an understanding of the organization, its environment and behavior.

When you change from one job to another you must deal with the unknown. Previously, you knew who was in charge of what, something about their methods and styles, and their position and responsibility, relative to the organization. In a new job you must discover all of the rules over again, although it may be easier than the first time, since you have experience and maturity on your side. To move from one job to a similar one is not too difficult. To change careers, however, even within the same field, requires more adjustments and more things to discover. This discovery requires seeing what you can do with what you have, as well as determining what areas you may need to develop and what new skills you may need to acquire.

## *JOB REQUIREMENTS/SKILLS*

Like so many other professional librarians, I too worked in the library before obtaining a Master's degree. Unlike many other professional librarians, I had advanced to professional level standing without the benefit of a Bachelor's degree. This career advancement was not without a considerable amount of work on my part which included extensive reading of the professional literature, visiting numerous libraries, attending conferences, workshops, and seminars, and learning from professionals both within the organization and outside. I wrote library reference publications and procedures manuals, participated in library professional groups, and eventually published and presented library science professional papers.

My work experience and the nature of the library and the organization for which I worked caused me to appreciate the skills and needs required for the knowledge worker in the future information age, although at the time it is unlikely that I would have thought of it in those terms. In any event, I began studies for a Bachelor of Science degree in Business and Management with a concentration in Technology and Management. Besides typical courses, such as accounting, economics, and management, it also included personnel management, organizational behavior, business and government, computers and data processing, systems analysis and design, systems management, and systems performance.

My Master's degree was also somewhat unusual in that most of the required core of library science courses was waived, based on my years of experience and demonstrated professional capability, both in my assignments and publications. This allowed me to concentrate on courses that dealt with information rather than librarianship, since I viewed information to be the fundamental cornerstone of the information professional, whether librarian or some other information-related professional.

Do not think, however, that library basics were omitted as a requirement for the degree; they were not. And, in fact, I took courses, such as "History of the Book," to round-out my background in areas where I felt I may be deficient in the more traditional coursework. Additionally, the degree program required two days of comprehensive examinations, relating primarily to library science core courses. Other courses completed included, "Introduction to Information Resource Management," "Managing Information Resources in Large Organizations," "Records Management," "Computer-based Information Retrieval," "Small-scale Computers for Information Handling," "Introduction to Computers and Information Processing," and "Information Retrieval Systems Design," for example.

The program was flexible enough to allow me to do independent research that pertained to restricted and non-restricted government information and commercial data, the findings of which were published.[2] This research included restricted information handling—military and commercial; regulations and methods for handling military and commercial data—including the use of computers and other technology; government data and the public versus private viewpoints; legislation and procedures; and the Freedom of Information Act, uses and future trends.

## *INFORMATION RESOURCES MANAGEMENT DEFINITION*

There are several definitions of what Information Resources Management (IRM) is. The two primary definitions, somewhat diverse, are: (a) that IRM is the management of all of the resources of an enterprise which are devoted to handling information; and (b) that information, itself, is a basic resource of an enterprise—a resource which must be managed. The first definition includes people, equipment, and procedures necessary to provide information to decision makers in the organization. The second definition of IRM says it is the management of the information as a resource and not the management of the resources involved in handling or producing information.[3] Either definition can be used for purposes of discussing the librarian's role in broader information-related careers. Here, however, in this case study the view of managing all the resources involved has been taken.

## *A CASE STUDY*

As Information Resources Manager for the Administrative Services Department, I serve as the central point in the Department Office for coordinating computer activities with the following duties: assist in conversion of manual systems to computer-based operations; formulate and write department computing plans; coordinate the development of department databases, including requirements plans; control and authorize all requests for computer hardware and software; ensure appropriate training for department staff in computer operations; and serve as the focal point for computer-related questions, problems, and requirements. A total look at systems and their requirements must always be considered to assure compatibility. Additionally, when requested, I review and prepare commentary on projects or problems, and serve as consultant to other departments on information-related projects, including policy, computer software and hardware, and associated functions.

Such activities require typical management skills, for example, planning, programming, budgeting, controlling operations, accounting and auditing, and evaluating. Further, it assumes that information has five stages in its life cycle: (1) requirements planning (determining what information is needed); (2) collection; (3) processing; (4) dissemination; and (5) disposition.

In carrying out this assignment I organized a department Personal Computer Users Group (ADO PCUG) of some 40 people. As chair, I schedule meetings, plan the agenda, and handle other related activities, e.g., demonstrations, overviews of products and methods, and discussions on computing. I also established the ADO

Computing Team, a subgroup of ADO PCUG. Each of the dozen members of this group has a specific task assignment, such as security, system configuration, a particular product for use in word processing, spreadsheets, database management—on both the personal computer and the mainframe—and so forth.

Additionally, I represent the Administrative Services Department on the Laboratory-wide Computer Systems Technical Committee (CSTC) and am a member of its Database Requirements Task Group, formed to oversee the implementation of Laboratory-wide databases. Furthermore as a member of the CSTC I perform as the communications link between Laboratory computing plans and ADO, and vice versa, and report on major events in both arenas. I also serve as representative to a Laboratory-wide central computing facilities users group named the Data Processing Branch Users Group (DBUG). As such I serve as the communications link between DBUG and ADO and report to DBUG on requirements, problems, and questions from ADO department users of the central computing facility.

In my present assignment I am called upon to participate in special task groups, such as personal computer maintenance, including evaluation of proposals; database management systems mainframe software; mainframe operating system tools and philosophy; the database management systems development project of the information systems group, concerning the design and development of Laboratory-wide databases, and especially those that are the responsibility of ADO; and text management mainframe software—evaluation and selection.

As Information Resources Manager for the Administrative Services Department, I operate as a facilitator and coordinator and as such I am involved in the conversion of manual, labor-intensive operations, to easier or less difficult automated systems for data handling. My role is to perform as a change agent within the department and the laboratory, to ensure a harmonious transition from paper to electronic format. At times the skills of a mediator are also required to effect an agreement or compromise. The mission is for the staff to increase their productivity and to gain in efficiency of operations, using the best tools available to them and at the same time to realize their own success and advancement, while improving the administrative support provided the technical departments.

Besides an understanding of the technological aspects of handling information one must understand the transition of data to information, and information to the end users. Data are simply data. Information is data transformed into information, based on the end user's requirements. Thus, it is imperative to understand the nature of the data with which you are dealing and the environment of that data relative to its intended audience.

It is evident that librarians, through years of service oriented functions, have a full understanding of this principle. Furthermore, librarians are trained in the five stages of the information life cycle. They know through training and experience how to determine information requirements, how to collect, organize, and present it, how to disseminate information, and to handle its disposition, relative to the clientele they serve.

The other players in this arena include the programmers and developers of applications, computing systems staff, and the administrators; all look from their own perspective. The librarian is trained to see from the side of the end user, irrespective of the type of information involved. Typically, programmers look for clean solutions in the design of their programs, systems people look for the technical challenge or feasibility of doing something, and administrators see as managers, often relying on the adages that, "it always worked before, so why change it," or, "if it ain't broke, don't fix it." That is not to say that each of these views is not equally important. The critical element is that different perspectives must be included in the design and operation of electronic systems to ensure the best solution for handling the requirements of the end users.

The profession of Information Resources Manager is an exciting one, but it is not without problems. For example the Information Resources Manager may become identified with the information problems, whether a software problem, a hardware problem, or simply an answer to a question not easily retrieved. This is especially true if the end user, either information seeker or information maintainer, relies on electronic data that you helped to assimilate. There is the tendency to blame something or someone else for not having or being able to produce the answer to a question or provide the solution. Previously, the same phenomena existed but the only solution available was the working files and skills of a particular person or group. The introduction of electronic data and systems suggest to many that such databases should perform miracles, not just what they have been designed to handle, often based on the same data and procedures as were previously used in the manual operation.

As we expand data processing into distributed functions there is the ever-increasing problem of determining the correctness of data. I see this as one of the most serious problems of managing information, whether for personal, corporate, or national use. Data integrity is critical to electronic operations, especially shared data. Librarians are well-versed in methods and operations used to match and retrieve data. They are typically confronted with information problems, see them as retrieval questions, and seek out the solution or answer as collectors and organizers of data in all possible formats. For the librarian there is the realization that data are presented in many forms to various groups for different uses. The librarian's proficiency is in recognizing this phenomena and ensuring that multiple end users will be able to retrieve that which they need, dependent on their personal requirements at various points in time.

## REFERENCES

1. Cortez, Edwin M.; Developments in special library education: implications for the present and future. *Special Libraries*. 77(4): 198-206; 1986 Fall.

2. Mount, Ellis; Newman, Wilda B.; *Top secret/trade secret: accessing and safeguarding restricted information*. New York; Neal-Schuman Publishers, Inc.; 1985. 214p.

3. Mendenhall, Gerald; Cook, Craig M.; *Information resource management for the information systems executive*. Arthur Young and Company; 1979 June.

# Research Scientist: A Career in Scientific and Technical Information Service

Robert E. Stobaugh

**SUMMARY.** At Chemical Abstracts Service, research is one of the areas available as a career. The background needed for research scientist positions may consist of chemistry or a related science, information science, computer science, or human factors, or various combinations of these. Present research involves expert systems, end-user interaction, natural language processing, chemical information manipulation, and other areas, for both internal and external purposes. Research offers a career that can be satisfying in the long run, and offers involvement in the future of the organization and of information science.

The word "research" brings to mind test tubes, extraction apparatus, elaborate balances, white mice, corn seedlings, experimental subjects, and many more ideas all involved with sciences such as chemistry, physics, biology, and psychology. But it is also essential in the field of information science and services. The environment may or may not be a laboratory, and the tools used are pencil and paper, calculators or computers and computer programs.

---

Robert E. Stobaugh is Manager of the Research Department in the Information Systems Organization of Chemical Abstracts Service (CAS), P.O. Box 3012, Columbus, OH 43210, where he is responsible for research and exploratory projects. Prior to his present position, he was Department Head of the Organic Abstract Editorial Department and Technical Advisor to the CAS Chemical Registry System, where he was involved in the initial development and operation of the Registry System. Dr. Stobaugh received a BS in chemistry from Rhodes College and the MS and PhD in chemistry from the University of Tennessee.

 

## *RESEARCH AND RESEARCHERS IN INFORMATION SERVICES*

Information services have always been familiar to many people, in the form of printed journals, magazines, newsletters, abstracts, and handbooks, for example. During the last decade or so, online computer-based information services have appeared and are being used more and more. Just as research plays an essential role in the development of new medicines or new audiovisual equipment, it is also very important in the development of theories and techniques for manipulating information.

I use the plain term "research" without any qualification (basic, applied, directed, free, process, product. . . ) to avoid endless and probably unproductive debate. And I also use a very simplistic definition, that of a scholarly or scientific investigation. Research may be carried out to discover new facts, to revise accepted conclusions in the light of newly discovered facts, or to apply new or revised conclusions.

The results of a research project may be such that no further work is recommended. Some reasons for not pursuing a line of research include: (1) the original premise or hypothesis may not have been proven; (2) the questions posed may not have been answered; (3) new facts may have been uncovered, but further development is not indicated because of technology limitations. However, research may also lead to conclusions which indicate a need for development projects that result in production systems.

At Chemical Abstracts Service (CAS), research scientists are recruited both by internal transfer from other parts of the organization, such as editorial areas (our editorial functions are concerned with document acquisitions and selection, abstracting, and indexing) or systems development, and by external recruiting. They usually have advanced degrees in chemistry or computer science. Many of those hired from outside CAS come from academic environments, where they have already been carrying out research. The author himself had thirteen years of production experience, ten in editorial activities and three with the Chemical Registry, before moving into the research area.

What do research scientists do at the office all day? Whatever the current project, whether voice input of data, investigation of chemical reaction information storage and retrieval, detection/correction of spelling errors, or analysis of information needs of chemists, all the work requires thinking, analyzing, creating and relating seemingly dissimilar things. There is nothing new or different here, since research has always required such activities. Only the tools have changed with computers and programs very much in use now.

## *RESEARCH AT CHEMICAL ABSTRACTS SERVICE*

To understand the significance of the research function, it is helpful to have an overview of CAS and its services.

CAS, a division of the American Chemical Society, is a non-profit information service organization which began operations in 1907 as an abstract journal. CAS collects all the literature published worldwide pertaining to chemistry and related fields, and then produces abstracts of the literature. CAS then prepares printed publications and computerized files for distribution.

To build this information base, CAS staff monitor scientific journals from throughout the world and select appropriate documents or articles for abstracting and indexing. Additionally, information is selected from patents, books, and symposia. As a result, chemistry-related information (e.g., author name, title, patent number . . . ) from about 450,000 documents per year is entered into our information base. Following the selection operation, technical staff examine the original document to prepare a brief summary or abstract of its contents. Also, a comprehensive set of index entries for each document is prepared. All information is entered into our database in English, no matter what the language of origin.

To give an idea of the operation's magnitude, the indexes prepared for the documents processed at CAS from 1977 through 1981 required 131,000 pages of index data, which contain 25 million index entries, and could occupy more than 11 feet of shelf space. The printed abstract issues for this same period could occupy about another 27 feet of shelf space.

In 1965, after 58 years of manual operations, we began to computerize the compilations of our main files. During the next ten years, various systems were built and enhanced so that by the middle of 1975 all information published by CAS was totally processed by computer with supporting programmed edits and manual edits in an online environment.

Since data are being processed through our computer system,

they can also be used in computer search systems. CAS licenses files to be used in in-house search systems. These files are also accessible via commercial online search services such as DIALOG and Questel. In addition, as CAS ONLINE, CAS files are searchable through our online international scientific and technical information network, STN International.

In 1965 one of the Departments in the newly established R&D Division at CAS was Chemical Information Procedures. Its purpose was to analyze the information flows and procedures at CAS associated with the processing of chemical information. In 1970 that department was retitled Chemical Information Science and continued in its analysis of chemical information processing, procedures, and work flow at CAS. In 1972, a research program was approved which was made a part of the Chemical Information Science Department. In 1978, to emphasize that CAS was processing more than just chemical information, the title of the department was changed to Information Science. In the early 1980s, CAS began to place increased emphasis on developing online services and allocated more resources to support contract commitments. As a result, during 1983 to 1985, research efforts at CAS were minimized in order to focus on the development efforts. However, starting in 1986, CAS made a commitment to re-emphasize the importance and the commitment to research at CAS. To emphasize the point, the title of this department was changed in 1986 to the Research Department.

Today, there are a number of key motivation factors driving the research program at CAS. Since the beginning of computer-based services, we have recognized the continuing importance to CAS of the information intermediaries who service end-users. We are now experiencing an evolution in the use of our services. In particular, as a result of the increase of online searching coupled with increases in end-user searching, CAS customer base and user community is going through a change which CAS is now addressing. In the technical arena, a number of significant transitions are occurring. More and more customer sites are placing computers in the hands of the end-user chemists. This, combined with advances in computer technology, leads to opportunities for CAS to place software and support capabilities direct in the hands of the research chemists and make access to CAS's databases and tools easier for the end-user. We do, however, expect that the role of the intermediaries will continue to be important.

In addition to the opportunities associated with user changes and technical environment changes, CAS itself is experiencing a number of processing changes in our internal environment. Because of increases in the amount of information which CAS wishes to glean from the chemical information that it processes, there is a growing cost in the data preparation area. Specifically, this is occurring with the processing of the chemical reaction, Markush, and new patent information which CAS is interested in providing to users. As a result, there are opportunities for efficiency improvements and significant cost savings by addressing the large dollar base associated with information gathering at CAS.

In looking to the future, our research is associated with foundation capabilities. These capabilities are necessary to gain an understanding of, and expertise in, major/new leading edge technologies or application of technologies to allow us to apply this in more targeted research in the future. Specific examples of these foundation capabilities are: artificial intelligence (including expert systems for problem solving and question and answering capabilities), natural language processing (which has potential in the areas of search query input, machine translation, specialized database building, and thesaurus capabilities), and human factors/cognitive psychology (which will be applicable to our online environments, and user interaction associated with our computer based systems). A fourth foundation capability is the area of computer science, that is, applications of new concepts in technology to the development of computer software.

Given the background and understanding of these basic foundation capabilities, together with other skills, CAS will perform research aimed toward products or processing, and exploratory areas.

Some research will be carried out with the intention to improve existing services, or establishing new services or products. Examples of such research are enhancements to our Chemical Registry System, organic synthesis, molecular modeling, and biotechnology. Other research will involve internal processing such as voice input of data, machine translation, automated abstracting and (where appropriate) the application of expert systems.

Exploratory research will address the investigation of novel concepts in the areas associated with information processing at CAS. An example of a research effort in this area is one which investigates ways of representing and organizing, in a computer system, the information content associated with chemical data.

## *REQUIREMENTS OF RESEARCHERS*

Research scientists in the past at CAS needed all of the following: chemistry background, knowledge of internal processes or products, and computer/information science. They all had advanced academic degrees in chemistry or chemical engineering, occasionally a degree also in computer/information science, and acquired CAS knowledge by experience. To a considerable extent this is still the case today, but research thrusts now into artificial intelligence and human factor areas have caused some changes. We now also look for candidates with backgrounds in psychology and in such areas as expert systems and natural language processing.

[2]Research experience in information or computer science would be useful, but is not required since few candidates have actually had such experience. But laboratory research, whether chemical, biological, psychological, or other, is helpful in that basic patterns of research are followed in any scientific investigation.

An appropriate ending to this article is the section that Dolan[1] used to begin an article on careers in online: ''Can you tolerate uncertainty, thrive on unrelenting change, uncover trends in an industry, and bring order to chaos? If so, you may wish to consider a career in planning or research and development.'' Such as these play a major role in the makeup of an information research scientist.

## *SUMMARY*

Research has become firmly established as one of the career possibilities in the area of information science. As online services continue to extend in coverage and usage, more personnel will be needed in this vital area. A position in research can be very satisfying—over the long run. Day by day, it may be a job without much structure and without visible gain for weeks or months. The rewards however are involvement in the future of the organization and even in the advance of information science, pride in a quality plan or product, high visibility, and independence.

---

[1]Dolan, Donna R. Careers in Online . . . Fourth in a Series Planning and Research & Development Jobs in the Online Industry. *Online*, 8(4):28-33; 1984 July.

# Online Database Manager: A Career in Scientific and Technical Information Service

Taissa T. Kusma

**SUMMARY.** Describes the duties of managing databases for the American Mathematical Society as well as the rewards, the requirements and the outlook for data base managers. Marketing responsibilities are discussed as well as the technical aspects of this sort of position.

## *INTRODUCTION*

After moving from Austria to Canada at the age of 13, I went to high school in Edmonton and Toronto and considered pursuing a career in chemistry or in engineering. My friends, however, advised me that engineering was not suitable for a girl. This made it sound even more challenging and I enrolled in the program at the University of Toronto and graduated four years later as the only girl in a class of 65 Chemical Engineers.

These were very good times—we all had multiple job offers even before graduation. I accepted a position at the Hercules Co. Research Center in Wilmington, Delaware, to work as an indexer in the Technical Information Division. I wrote abstracts of company research reports and indexed them for the card catalog. In addition to indexing reports, I selected articles from the journal literature and wrote abstracts for the bi-weekly current awareness bulletin. The Technical Information Division was separate from the library, the indexers were all subject specialists and were called literature

---

Taissa T. Kusma is Manager of Database Services at the American Mathematical Society, P.O. Box 6248, Providence, RI 02940. She received the BS in Chemical Engineering from the University of Toronto and the MLS degree from the University of Rhode Island.

 

chemists. The head of the Division was Dr. Herman Skolnik, the well-known pioneer in chemical information retrieval. Even though it had been my intent to stay only one year, Hercules was such a good place to work in, that I stayed for 3 years, then married and moved to New Jersey where my husband was opening a pediatrics practice.

My next job was as head of the chemical research library at Merck, Sharp & Dohme in Rahway, NJ, where I worked for Paul Stecher, the editor of the Merck Index. It was an interesting challenge, running the library without formal library training. One of my projects was the re-cataloging of the library collection from the Dewey Decimal to the Library of Congress Classification system. Less than 2 years later I left and moved to Rhode Island.

After more than a decade spent as wife and mother, I returned to school and received my master's degree in library science from the University of Rhode Island in 1975. Several years and several rather uninspiring library jobs later, I was offered the position of Database Specialist for the American Mathematical Society (AMS), to develop and manage MATHFILE, the online version of Mathematical Reviews (MR). The requirements were a degree in mathematics or another physical science, knowledge of other languages and MLS degree. Familiarity with online systems would be helpful but not essential. I accepted the position even though considerable travel would be required and I had a terrible phobia of flying.

I began my new career at the AMS on July 6, 1981—a career very different from the traditional librarianship as I knew it. Even though my library degree provided a good foundation, most of the work I was about to undertake had to be learned on the job. My library school training had barely touched the field of databases and online retrieval, although I had at least taken several l-day training sessions from the major online vendors.

There was so much to learn! I had gone from a library environment into the publishing world: The Society is a publisher of books, primary journals, and the prestigious abstracting/reviewing service, Mathematical Reviews. As a librarian, I had dealt with finished publications, now I was working on producing one (MATHFILE User's Guide) and turning another into an online database. The publishing business could be learned from within the AMS, but databases and the online industry had to be learned at meetings and from colleagues. I attended meetings such as Online, ASIS, SLA, ASIDIC, and got to know the people with jobs similar to mine. I was to realize later that this network of contacts in the field would become my most important and valuable resource.

## *THE JOB OF DATABASE PRODUCER*

Developing and managing MATHFILE (later renamed Math/Sci) became the most challenging job of my life. Much was learned through trial and error: the design and development of the file, dealing with vendors and developing a market for the online file, persuading the mathematical community that Math/Sci is a valuable research tool that could save them countless hours of searching, telling librarians that the file contained more than mathematics, and covered other areas such as statistics, computer science, econometrics, linguistics, and others. Since I was the whole Math/Sci department, it was quite a feat to travel around the country giving workshops and at the same time mind the business at the office. During the second year I gave close to 40 training sessions in the U.S. and Canada.

My job could be divided broadly into marketing and database development. Marketing is a term that encompasses everything outside the technical work on the database. Promotion (including exhibits); brochures, mailings, advertising; user services (including training, documentation, directory listings, help desk); contract negotiations with vendors and other publishers; pricing the file; planning new products and additions to the file. Database development includes designing the online file that is sent to the vendors. This involves working with our systems people who produce and maintain the Mathematical Reviews files from which the printed publication is produced. Before the file can be loaded on the vendor's online system the right output format must be worked out with the vendor and tested in an online testfile. This process is repeated with each of the three vendors who load Math/Sci. Update tapes are prepared monthly and sent to vendors. Periodic checking through online searches is done for quality control, and vendors are notified should a problem arise. Often users are the first to discover if anything goes wrong and they call us or the vendors directly to report the problem.

## *MARKETING*

A. *Contract negotiations with vendors*. Each vendor has a standard contract for acquiring databases; most producers have their own contracts for leasing their files. Each side wishes to have its own contract signed, and so a mutually acceptable agreement must be negotiated. Annual agreements are renewed, often with changes in royalties and sometimes in other clauses. Various terms are contested in negotiations: the vendors seek to obtain exclusive or multi-annual contracts; producers try to gain from vendors information about their users, improvements in file design, a reload of the file, or speedier processing of updates.

B. *Training*. Onsite user training is an important and time-consuming part of the marketing program. Planning and organizing the training program involves finding local hosts willing to sponsor the sessions. The host can be a local academic library, an online user group, or the local chapter of SLA or ASIS. Arrangements must be made, publicity arranged, and several sessions organized for each trip to minimize travel expenses. Sessions are most often given at academic libraries, most are 1/2 day long, others can be one-day sessions. A theoretical presentation is followed by hands-on practice session.

In addition to workshops for librarians, demonstrations are often held for the faculty and students of the departments of mathematics, statistics, and computer science. Whenever possible, I hold training sessions before or after a Math/Sci exhibit to save on travel costs. This is not always easy since the time and place may not be convenient for the local hosts.

C. *Documentation*. Database specifications are written and given to vendors so that the file can be properly designed and processed before loading. User documentation is prepared as aid to effective searching. The first MATHFILE User's Guide was published by AMS in 1982. Since then many changes have occurred in the file, including a name change, and we have just published a new Math/Sci User Guide. Most database guides are issued in a loose-leaf binder for easy updating. In the case of Math/Sci enough changes have taken place to necessitate a completely new edition.

As editor of the Guide, I am responsible for putting the publication together, and for writing the chapters which describe the file, its printed equivalents, and various search techniques. Work on other sections is done by the MR editors, the MR librarians, and the

Systems Group. Included are lists of serials covered by the database, the subject classification schemes, and lists of descriptors and mnemonics used to represent mathematical symbols.

Other types of documentation are written periodically, such as descriptions of the file and sample searches (these are rather time-consuming) for vendor/user update meetings; questionnaires are filled out for database directories; transparencies are prepared for training workshops and for product review sessions at meetings; exhibit descriptions are written for meeting programs.

D. *Promotion*. Many different activities come under the heading of promotion.

1. Exhibits at library, online, and professional meetings. Running exhibits is complex and requires much time and work, but is essential and can be very rewarding and interesting. The work consists of planning the schedule of 8 to 10 exhibits a year and making all the arrangements: filling out order forms and processing payment for booth space, furniture, telephone service, electricity, computer equipment (or arranging to have it shipped from the office); reserving the time for a product review, sending an abstract and a description of the exhibit for the program, and making hotel and travel reservations. The supplies for the exhibit have to be prepared, packed and shipped in advance, such as boxes with booth panels, printed promotional materials and sample publications. Until recently, we shipped our microcomputer equipment, but after several cases of damage in transit we are now renting on site. Whichever method is used, hardware problems can turn up unexpectedly and spoil an exhibit. The work at the booth can be exhausting, but is very interesting—the busier the meeting, the less tiring and more rewarding the work. Whenever possible, I try to arrange one or more workshops during the same trip to save time and travel costs.
2. Brochures. A general brochure describes the content of the database, its strengths, applications, and availability. I write the text and work with a designer on the final format, the colors and other details of presentation. Printing is done in the AMS printshop and sometimes by outside printers. Brochures are used in exhibits, mailings, and in response to requests for information. We have two types of Math/Sci brochures—one with general information about the database, the other for

end-users, explaining the benefits of online information retrieval.

3. Help Desk. Users with questions about the database call our 800 number; some seek information, others ask for help with searches. Such calls tend to come in clusters, with some days busier than others. Providing telephone search assistance is one of my favorite jobs. Sometimes a call will interrupt some urgent work, but each caller quickly gets top priority and all my attention.

## *DATABASE DEVELOPMENT*

The database MATHFILE was first produced as a by-product of the printed publication, *Mathematical Reviews*. The reviews and abstracts were keyboarded with typesetting codes for the printer. Before online tapes could be prepared for the vendors, these codes had to be removed from the file. Mathematical expressions in the abstracts were also encoded and had to be changed to mnemonics that could be easily read and deciphered by the online users. To do the required code conversion, a computer program was developed by AMS and was named the pH7 program. Neutralizing typesetting codes was easy enough, but pH7 was also expected to transform some complex mathematical expressions to a simple series of mnemonics, from which the user could reconstruct the original mathematical expressions. It worked most of the time, but in the more complex cases, pH7 sometimes neutralized essential information, bringing about an irreversible reaction! Worse yet, some characters used in the conversion interfered with the vendor's software and required extensive corrective work by both sides. Clearly, a better system was needed if the online database was to be fully transportable, mathematics and all.

## *TeX SOFTWARE*

The needed improvements came in 1985, when the AMS changed its typesetting system for Mathematical Reviews. It adopted TeX, a superior computer composition system developed by Donald Knuth of Stanford University. TeX was especially designed for the production of technical books containing special characters, such as mathematical symbols. TeX allows faithful reproduction of the most complex mathematical expressions without loss of accuracy.

The breakthrough for online application of TeX came in 1986 when software packages for the PC and Macintosh microcomputers were developed with a full implementation of the TeX system. The software allowed users to obtain online output in high-quality typeset form with actual mathematical symbols. Ordinary dot-matrix printers could now produce publication-quality prints of online records and, as additional software was made available, the output could also be displayed on the screen! Authors were quick to adopt TeX as the best desk-top publishing system for technical books and papers. It was an exciting time for Math/Sci, as it became the one database whose online output could be typeset directly on the PC. However, there was a price to be paid for these enhancements. The file was now different and had to be redesigned by the vendors before it could be loaded. A long interruption in Math/Sci updates resulted.

Graphics had become the star attraction at Math/Sci exhibits and users were expressing interest in buying the software. We decided to make TeX software for micros available through the AMS, primarily as an enhancement to Math/Sci. Now we had two products to market: the database and the software.

## *EXPANSION OF MATHFILE INTO Math/Sci*

The name MATHFILE was changed to Math/Sci with the addition of three new subfiles: CMP (Current Mathematical Publications) a current awareness subfile and two statistics subfiles: CIS (Current Index to Statistics) produced by the American Statistical Association (ASA) and Institute of Mathematical Statistics (IMS) and an older statistics file, Index to Statistics and Probability (1902-1968) by John Tukey and Ian Ross. These additions brought much additional work in file development and in contract negotiations with our new partners. The work became ever more challenging and exciting, and the time in ever shorter supply. Hours of work were extended into evenings and weekends. A terminal in my bedroom, linking me by phone with our office mainframe computer, allowed me to work late into the night. I was still a one-man operation (except for our database production group), but the work load had expanded greatly. There was also increased participation in professional organizations, such as ASIDIC and NFAIS. Time available for home-related activities shrivelled, leisure time disappeared quickly. Vacation days remained unused and spilled over into sick-leave, fortunately also unused.

An interesting result of this time shortage was a tremendous increase in the intensity of experience during leisure moments, such as driving to and from work, and seeing a beautiful view or a sunset. As a scarce commodity, each moment became precious, to be enjoyed to the fullest. As a means of self-preservation, I took up race-walking for pleasure and coincidentally for exercise. During the summer, I would leave work around 6:30 and arrive at the State park before sunset, walk a 2.5-mile loop around the lake in 33 minutes, enjoy a 10-minute leisurely swim and the beautiful view of the twilight sky, and return home 5 minutes later. The whole trip took just over an hour and gave me a daily sustaining dose of enjoyment, relaxation, vitamins, and exercise all in one package. This allowed me to work often up to 16 hours a day, 7 days a week, with this one hour a day as my reward.

## *JOB REWARDS*

My job is unusually interesting and challenging. It has given me the freedom and opportunity to develop, to learn, to do and achieve as much as my resources and the available time will allow. The job is often exhilarating in the amount of responsibility and the diversity of functions it provides. In return, I give it my total involvement, all my time and energy.

The diversity of responsibilities in my job is partly due to the relatively small size of our operation. In larger organizations with larger staff, the positions are more narrowly defined and each person is responsible for a smaller part of the overall database operation, but at the same time can rely on more support staff. I much prefer the variety of managing many functions in a smaller operation.

However, the work load has become too heavy for one person to handle, even with the long hours. Fortunately, I now have an assistant, a librarian with a knack for microcomputers—exactly what I needed. With his help, I hope to reclaim some of my weekends and evenings for non-database activities. On second thought now we can finally turn to the business of writing the newsletter: "Notes from Math/Sci."

## *ONLINE CAREERS*

The online industry holds many opportunities for librarians who wish to go outside the library and become the providers of services to libraries. This field is best suited for the person who enjoys travel, interacting with and serving people, variety, and the challenge of dealing with the new and unexpected. The field is still new—exciting technical developments and new applications are taking place continually and career possibilities for information specialists are expanding.

# Abstractor and Indexer: Careers in Scientific and Technical Information Service

Ellen Young
Elliott Linder

**SUMMARY.** Information science can be a professionally satisfying alternative to industrial or academic careers at the bench for the technically trained individual. Abstracting and indexing serve the technical community directly, saving time and effort in acquiring information from the literature and patents. Descriptions are given of the abstracting and indexing operations at the Central Abstracting and Indexing Service of the American Petroleum Institute with emphasis on the information professionals involved.

## *PART I: ABSTRACTING AT CAIS*

***Ellen Young***

### *Role of Abstractors*

"Those who can, do; those who can't, teach." Extrapolation of this attitude implies that a career in information science is less desirable than a career in the laboratory or on the administrative staff of a technology company.

But a closer look shows that personal job satisfaction in an information service can be high. In a service covering a variety of subject matter, a literature abstractor will obtain a wide overview of technology that may not be possible for an industrial chemist focussed on more specific problems. Great depth of field is another advantage; for example, coverage of industrial catalysis will em-

---

Ellen Young is Senior Abstract Editor, and Elliott Linder is Senior Editor, Technical Indexes, at the Central Abstracting and Indexing Service of the American Petroleum Institute, 156 William Street, New York, NY 10038. Ms. Young holds a BA in Chemistry from Hunter College. Mr. Linder holds a BS ChE from New York University.

 

brace many of the leading catalysis journals and material ranging from the practical to the highly theoretical. For abstractors, another plus is the ability to use foreign language skills in abstracting from journals published in German, French, Russian, Spanish and other languages. And then there is the problem of maintaining interest when abstracting. The perils of falling into error, of being a sort of high-wire artist, help prevent the work from becoming routine. Like the art of ballet, abstracting becomes possible, but never easy.

Professional satisfaction is another plus. The information scientist aids directly in extracting knowledge from the literature and disseminating it to the interested community. By weeding out material not of interest and condensing the relevant information, the information scientist also saves considerable time for the technologist.

At the Central Abstracting Service of the American Petroleum Institute, the literature abstractor selects the articles to be abstracted from a list of journals and meeting papers set up by advisory committees from the petroleum industry. The subject matter of the Refining, Health and Environment, Petroleum Substitutes, Transportation and Storage, and Oilfield Chemicals bulletins published includes, for example, petroleum processing, unit operations, catalysis, petrochemicals, safety, petroleum substitutes such as shale oil and tar sands, health and environmental concerns related to petroleum and its products, and chemicals for enhanced oil recovery. The abstractor chooses material ranging from business and economic news to highly technical material in chemistry and chemical engineering. A written set of selection rules and guidance by the literature bulletin editors aid the abstractor in making the choices.

To ensure style conformity, the abstractor follows the API Abstractors' Manual with respect to style and format, rules for spelling of certain words, compounding of words, abbreviations of technical units and other expressions, transliteration of foreign language titles, formatting of literature references, and the bibliographic information supplied with each abstract. Natural language is used.

Whenever possible, the abstractor writes an informative and direct abstract, rather than a descriptive or indirect text. The statements are exact, concise and unambiguous. The text is selective, but not critical; information pertinent to the petroleum industry is emphasized and other details are minimized. New data are stressed. Exact figures, rather than general statements for experimental conditions and results are given whenever feasible. The abstractor writes a descriptive abstract when an article, such as a review, contains too much information to be abstracted concisely, when it is a discussion of known information, or when it contains mathematics difficult to verbalize. Unexplained acronyms and abbreviations are avoided whenever possible.

Preferably, the main point, or results, of a study appear in the first sentence of the abstract and other details follow, as in the following example:

> **A new mathematical model relating asphalt viscosity to penetration is applicable to road asphalts and asphaltic cementswithin the 25-240 penetration range. The model was derived by applying a nonlinear polynominal regression program to 90 published experimental data points for different grades of asphalt. The deviation between experimental viscosities and those calculated with the new model for 90 asphalt samples did not exceed 10%. Tables, graph, and 10 references.**

As the final step, the abstractor assigns one or more "section headings" to the abstract. In each of the five literature bulletins, the abstracts are grouped under headings such as Catalytic Conversions, Motor Fuels, Water Pollution Control, Pipeline Descriptions and Plans, Coal Gasification, Synthesis Gas, Fracturing Fluids, and Economics and Statistics. The section headings classify the abstracts for the convenience of the bulletin reader and are also searchable online.

To avoid repeating information covered in a previous abstract, the abstractor may insert in the text a bibliographic reference covering the material previously abstracted. Microfiche files of previously abstracted short news items and meeting papers are checked to ensure that the news items are not done again and so that cross references to meeting papers can be provided, thus giving the reader another source for obtaining the paper.

### *Training of Abstractors*

One of the most important functions of the literature editors is training new abstractors. The editors begin by acquainting the new abstractor with the Abstractors' Manual, and some of the basic rules and procedures to be followed. They assign a short list of journals thought suited to the abstractor's technical background and interests, show him or her how to select from the journals, have the trainee abstract a few articles at a time, and immediately edit them so that the trainee will know very soon what errors in content and form he is making. All initial abstracting is from English language journals so that complications arising from handling foreign languages are not superimposed on other problems. A new abstractor is heavily edited for six months and followed closely for two years, and his or her article selections are checked.

The editors check each abstract for correctness of bibliographic information, good English, and abstract content and form. They edit new abstractors more carefully, and experienced abstractors, who are technically trustworthy, more lightly. The editors also check the accuracy of the section headings under which the abstract will appear in one or more of the literature bulletins.

One of the most important aids in reducing errors in content and section heading assignment is feedback. The indexing editors provide the most immediate information; as they edit the indexing of the abstracts, they bring such errors to the abstract editors' attention within a relatively short time after abstracting of a given document is complete. Feedback also comes from the readers of the literature bulletins and users of our online services; they may point out mistakes and suggest areas where interest is high and coverage should be broadened. Skimming of the bulletins by the editors is another useful feedback device. The chief editor then brings such errors to the attention of the abstractor/editor involved.

Another useful device for quality control is abstracting staff meetings at which errors, various trends in the literature, areas of high user interest, changes in procedures, and other matters can be brought to everyone's attention.

Abstracting production rates vary considerably, depending on whether a document is well or poorly organized, on the technical difficulty of the material, on the language if not in English, and to a lesser extent on the length of the document. An Abstractor working mostly on Russian, French, and English language material of high technical difficulty would be expected to do a minimum of 30-35 abstracts per week. A person with a load consisting largely of short news items and relatively non-technical, well-organized material, would be expected to do 40-45 or more a week.

Editorial production will also vary. The editing of cross-references to previously published information and of abstracts consisting of a title only, will require little time. Seasoned abstractors of high ability will be edited much more quickly than those of less ability or new abstractors. The same factors that affect abstracting speed can slow the editing of less able abstractors. An average editing speed of four abstracts per hour is thus expected.

The abstractors currently enter their abstracts directly onto a "dumb" computer terminal for transmission to a central minicomputer where the abstracts are stored and from which the editors receive printed copies of them. After editing of the hard copy is completed, the abstract bibliographic information is checked on a terminal by the copyreading group and released so that the stored abstracts are collected on tape and grouped for inclusion in the appropriate literature bulletins. Typesetting by computer and offset printing of the bulletins follow.

Very soon this text-editing system will be modified so that the abstractors will enter their work on a personal computer, and editing and copyreading will take place on personal computers without the intervention of printed copy.

After release, a printed copy of each abstract is passed to the indexing group, where it is indexed for eventual inclusion in the online file.

## *PART II: TECHNICAL INDEXING AT CAIS*

**_Elliott Linder_**

### *Introduction*

Since 1964 the Central Abstracting and Indexing Service (CAIS) of the American Petroleum Institute (API) has been producing two technical databases, APILIT and APIPAT. These two bibliographic databases cover a wide range of subject matter of interest to the petroleum, gas, and petrochemical industries: processing, analysis, chemistry, safety, transportation, substitutes, environmental problems, toxicity, legal issues, engineering, finished products, oilfield chemicals, etc.

In order to produce APILIT and APIPAT, a controlled vocabulary is used to index the abstracts of the covered articles and patents. It is to the system we have developed to accomplish this, to the tools which are employed, and, most of all, to the people who use them, the indexers and editors, that this article is addressed. For the sake of continuity with Part I, my discussion here will concern itself with material abstracted in-house at CAIS, that is, material which is indexed for the APILIT file. The indexing procedure for APIPAT is essentially the same.

### *Indexing and Editing Procedure*

As mentioned in Part I of this article, the abstracts are entered into a text-editing system which produces either weekly or monthly tapes used for computer typesetting of the abstract bulletins. When the abstract tapes are generated, two other things occur simultaneously: a computer printout of each abstract (Figure 1) is generated; and the computer bibliographic page for each abstract is passed into the indexing mode and assigned a blank indexing page.

The printed abstracts are given to an index editor, who distributes them to the indexing staff according to indexers' area of expertise. One indexer may specialize in chemistry, another in engineering, another in coal conversion, for example. Of course, there may be considerable areas of overlap, and the editors are expected to be knowledgeable over the wide range of material covered.

#### *Indexing*

Upon selecting an item for indexing, the indexer first does a cursory reading of the abstract just to get an idea of what it is about. He then begins the process of assigning index terms which represent the bibliographic and subject information contained in the abstract. This is accomplished by using the API Thesaurus,[1] the main body of which consists of an alphabetical listing of the valid index terms—controlled vocabulary—and cross references thereto. It is the goal of the indexer to index each concept he encounters in the abstract, by the most specific term, or terms, available.

All of the indexing at CAIS is done on the text-editing system. At this time, each indexer has a "dumb" terminal which is connected to a minicomputer. In the near future, each will have a self-contained personal computer. During the indexing process, the system is used in the "INDEX ENTRY" mode.

In order to access the document he is about to index, the indexer

---

[1]*API Thesaurus*. Washington, DC: American Petroleum Institute, 1986.

287319 SEPT #39 33-06553 INDEXER: INDEX EDITOR:

F. E. Anderson; J. M. Prausnitz (Univ. Calif. Berkeley)

AIChE J. 32 #8:1321-33 (Aug. 1986)

INHIBITION OF GAS HYDRATES BY METHANOL. A molecular-thermodynamic correlation for calculating the amount of inhibitor that must be added to prevent hydrate formation in gas processing was developed and validated by comparison with a wide variety of published experimental data. It can be used in computer-aided design for gas processing and transportation by pipeline. The method calculates hydrate-point pressures and temperatures, and coexisting-phase fractions and compositions for gas/water/methanol mixtures at 220-320 K and up to 500 bar. Tables, graphs, and 43 references.

-NATURAL GAS, NAT. GASOL., LPG
-PIPELINE.OPERATING PROBLEMS
-OXYGEN COMPOUNDS

FIGURE 1. Computer Printout of an Abstract

keys in its temporary abstract number (TAN) on his terminal. He then transmits this number to the minicomputer, which returns to the screen the bibliographic page (Figure 2) for the document. He verifies the page and enters his initials in a space provided at the top of it. The biblio page is then transmitted, and the system furnishes a blank indexing page on which the indexer will enter the terms he selects.

The indexer enters each term selected into the working space of the index page on the screen of his terminal. He then directs the system to place the term on any one of the 42 lines available per page. When he does this, the term is transmitted to the minicomputer where it is verified against the computer dictionary and placed on the designated line, all within a matter of seconds. If the term entered is not valid, the system returns the term to the working space, blinking, for correction. The system also allows for replacement, deletion, or duplication of any term previously entered, and additional index pages may be requested.

Occasionally in the course of the indexing process, the indexer may encounter difficulty, either with the clarity or precision of the abstract or with understanding the subject matter itself, especially if it is a new or unfamiliar technology. In the first case, he is obliged to consult the original article, which is always available, for clarification. In the second case, he will normally seek help from one of the editors, whose responsibility it is to provide technical expertise to the staff and to answer questions regarding indexing policy.

When the indexer is satisfied that he has assigned all the terms necessary to express the information contained in the abstract as specifically as the vocabulary will allow, he checks over his work to be sure that any role indicators and links* have been assigned where required. Then he does a subject analysis of the abstract to ensure that terms representing the major subject of the document have been assigned. Oftentimes the subject of a document is not stated per se, especially not in the language of the Thesaurus, so this step is most important. It is to these major subject terms, whether indexed previously or at this stage, that the indexer assigns the "major term" designation by placing a special character—(* or P)—in front of the term on the screen in a space alloted for that purpose. Every

---

*The Role Indicators A and P are assigned to terms when they are agents or products, respectively, of chemical reactions. Links are assigned to establish a special relationship between two or more terms. Both features allow for greater specificity in the retrieval process.

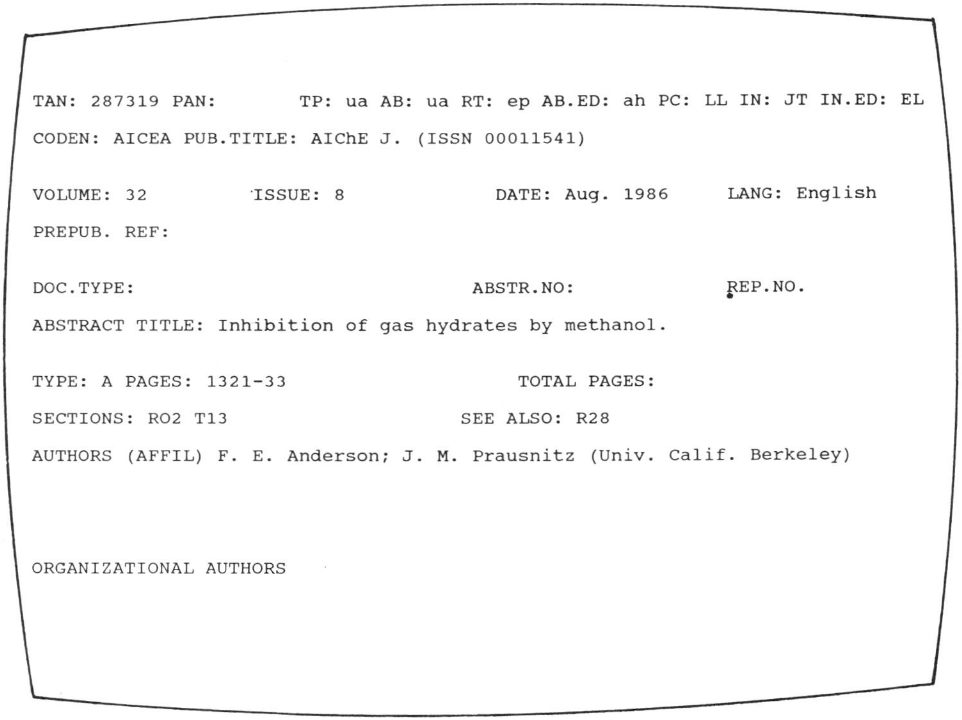

```
TAN: 287319 PAN:       TP: ua AB: ua RT: ep AB.ED: ah PC: LL IN: JT IN.ED: EL
CODEN: AICEA PUB.TITLE: AIChE J. (ISSN 00011541)

VOLUME: 32          ISSUE: 8             DATE: Aug. 1986     LANG: English
PREPUB. REF:

DOC.TYPE:                                ABSTR.NO:            REP.NO.
ABSTRACT TITLE: Inhibition of gas hydrates by methanol.

TYPE: A PAGES: 1321-33                        TOTAL PAGES:
SECTIONS: RO2 T13                        SEE ALSO: R28
AUTHORS (AFFIL) F. E. Anderson; J. M. Prausnitz (Univ. Calif. Berkeley)



ORGANIZATIONAL AUTHORS
```

FIGURE 2. Computer Bibliographic Page for Abstract of Figure 1

index must have at least one major term, and when the indexer, having finally completed his work with a document, transmits the entire index (Figure 3) to the minicomputer to await editing, the system checks that at least one major term is indeed present. This check completed, the system requests a new TAN.

On average, each indexer produces 20-25 indexes per day, but in actuality the number may vary between 10 and 50 depending on the degree of difficulty of the subject matter. As a rule, those documents which contain a great deal of complex chemical information take far longer to index than those which deal with, for example, economics or engineering.

Either during or at the end of the working day, the indexer places the computer printouts of the abstracts he has indexed in a drawer for that month's material. To some of the more complex abstracts, he may attach a computer-generated hard copy of the index, usually multipage, to enable the editor to get a clearer picture of the organization of the work without having to flip back and forth between pages on his terminal, where only one page at a time can be examined. It is from this drawer that the editors take material to work on each morning.

*Editing*

All editing at CAIS is done on the text-editing system as well. In the same fashion as the indexer, the editor accesses a desired document by keying in its TAN, this time in the "INDEX EDIT" mode. As before, the system responds with the bibliographic page, which is verified by the editor and onto which he now keys his initials. The system then furnishes the first page of the index completed by the indexer. Before proceeding to edit the index, the editor does his own cursory reading of the abstract to get a sense of its content.

It should be emphasized here that although in theory the goal of the indexing process has been to express all of the concepts in the abstract as specifically as possible, i.e., to create the perfect index, such a goal is elusive, at least partially subjective, and, in most cases, not economical. It is not the purpose of the editor to re-index the abstract by checking every entry made by the indexer. He operates under the assumption that, since the indexing has been done by a professional who has been thoroughly trained—perhaps even by the editor himself—the work will need only minor revision,

```
   ARL           -TERMS-                      ARL            -TERMS-

01 D   -NATURAL GAS, NAT. GASOL., LPG   22 D   MULTIPHASE
02 D   -PIPELINE OPERATING PROBLEMS     23 D   CONCENTRATION
03 D   -OXYGEN COMPOUNDS                24 D   MIXTURE
04 *   HYDRATE FORMATION                25 D   TEMPERATURE -100 TO -10 C
05 D   INHIBITION                       26 D   TEMPERATURE -10 TO 20 C
06 P B METHANOL                         27 D   TEMPERATURE 20 TO 40 C
07 P B HYDRATE INHIBITOR                28 D   PRESSURE 1500 PSIG AND HIGHER
08 D   GAS PROCESSING                   29
09                                      30
10 P C NATURAL GAS                      31 D C CARGO
11 D C WATER                            32
12 D C CLATHRATE                        33
13 D   PREVENTION                       34
14 N   DATA CORRELATION                 35
15 D   MATHEMATICS                      36
16                                      37
17 D   COMPUTER AIDED DESIGN            38
18 D   TRANSPORTATION                   39
19 D   PIPELINE                         40
20                                      41
21                                      42

TAN: 287319
```

FIGURE 3. Computer Index Page for Abstract of Figure 1

and proceeds according to that assumption. What he pays most attention to are the major subject terms: he wants to be sure that terms representing all the major concepts in the abstract have been indexed and designated as major terms; he also wants to be sure that no terms have been indexed which represent concepts not present in the abstract due to a misinterpretation by the indexer. If unsure of himself on occasion, he can always avail himself of the expertise of another of the editors. Two heads are often better than one.

### *Tape Creation and File Updating*

When the editor has completed making his changes and is satisfied that the index is of file quality, he releases the index to a special storage area of the minicomputer's disk to await monthly index tape generation, a process which takes place when all of a given month's material has been released. The product of the process, the monthly APILIT update, is forwarded to our vendor company where it is added to the cumulative APILIT file available for online searching. The APILIT file is currently available on the Systems Development Corp. (SDC) ORBIT system. As of March 1987, APILIT will be available also via the STN International system.

An editor reviews, on average, 40-50 indexes per day but, as with the indexing, the number may vary widely, between 20 and 100, depending on the length and complexity of the material.

## ***Quality Control***

CAIS takes several steps in order to ensure the quality, i.e., the accuracy and consistency, of its indexes. The first of these is the training of its staff.

### *Training the Indexer*

When a new indexer is hired by CAIS, he is placed under the tutelage of the Senior Editor, whose responsibility it is to guide his development into a fully productive indexer within a period of six months. The first few days are spent on familiarizing the trainee with the CAIS operation and on explaining the philosophy of the indexing system. He is also required to read the API Indexer's Manual, which describes the fundamentals of the indexing system and provides rigorous instruction on the use of the Thesaurus. After this, it is a matter of practice, practice, practice.

The trainee is supplied with a large variety of previously indexed titles on which to test his developing skills, starting with very simple material. Each item is discussed with the Senior Editor, who is at his disposal at all times. As his proficiency increases, more difficult material is supplied, and the formal training process continues until the trainee has progressed to the point where he can handle production material.

### *Editing and Feedback*

The learning process does not stop with the conclusion of the formal training period; quite the contrary. Since almost all of the indexing is reviewed by the editors, the editors are in an ideal position to provide continual critical feedback to the indexers on their completed work (as well as the technical expertise they provide, when called upon, during the indexing process). If, during the editing process, the editor finds that the indexer has either missed the boat or has been careless or incomplete in his work, he can do any or all of three things.

The first possibility is to discuss the item in question with the indexer directly. I find this to be the most positive approach because it affords for an active interchange between the indexer and editor. If the indexer is treated with respect, such a discussion usually yields positive results.

A second approach is for the editor to make written comments at appropriate places on the printed abstracts and to return such items to the indexer at the end of the day. This is satisfactory in cases where little explanation is necessary, i.e., where the errors are more obvious or clear cut.

Finally, in the rare case where extensive revision of the index will be necessary, the editor may choose to generate a printout of the unedited index upon which he indicates the corrections to be made. This is returned to the indexer and is usually accompanied by a discussion of the reasons for the changes.

### *Monthly Staff Meetings*

A further forum for feedback exists at our monthly staff meetings. On the final Tuesday of each month, the indexers and editors come together informally to discuss a variety of work-related issues.

There is no real agenda. The Senior Editor does present a list of problems he has encountered during the month, but discussion is by no means limited to that list. Everyone is encouraged to contribute.

It is at meetings such as these that problems common to the group may be exposed, even solved; that suggestions arise and decisions are taken which improve productivity; that vocabulary is clarified; that the way we index is uniformized; that a sense of community can be felt. It is at meetings such as these that part of the communication takes place which makes for continuity in the accuracy and, as importantly, the consistency of the indexes produced at CAIS.

## *The Indexer: Who Is He?*

### *Qualities and Prerequisites*

"It helps to have a good sense of humor." Such is my reply when I am asked about what qualities are desirable in an indexer. What I mean is not the obvious. What I mean is that the same sharpness, the same quick-wittedness required to appreciate or to make a good joke is a most desirable quality in an indexer. And it is that quality, combined with a good memory, which enables the indexer to perform one of his most important functions: concept recognition in the language of the Thesaurus.

Since a major part of the indexing process involves rigorous use of the Thesaurus, the indexer must be able to understand its structure and to follow the instructions for its use; this implies an ability to think in logical patterns. He must also be able to do subject analysis of a wide variety of material; this is facilitated by a good command of English. Because a significant proportion of the material we handle is chemical in nature, we require that each member of the technical indexing staff have a degree in chemistry or chemical engineering or equivalent technical experience. The ability to operate, or to learn to operate, a computer terminal is preferred.

### *Professional Satisfaction*

The job of technical indexer or editor at CAIS can yield professional satisfaction of various kinds. Not all chemists desire to work in the laboratory or the process plant, or in academia. There are those for whom the satisfaction is the exposure to a growing variety of technologies and research developments in the literature. For others it is logically organizing information and making it easily retrievable to the scientific community. (It is always nice to hear from a client that he got the best results searching your file because your indexing is the best.) For still others it is the opportunity to work with a sophisticated system which makes such logical organization of information possible.

For me, it is all these things. Yet, it is something else, something more: it is a fascination with language and vocabulary. It is the translation of the language of the literature into the language of the Thesaurus. And it is the opportunity to contribute to the development of the indexing vocabulary which provides the greatest satisfaction of all.

### *The Future: Machine-Aided Indexing*

Compared to using classification schemes and natural language systems, it is expensive to index using a controlled vocabulary; it is expensive to edit that indexing; and it is expensive to maintain and annually update the vocabulary of a thesaurus.

Fortunately, CAIS has embarked upon a project designed to reduce the cost of producing its technical databases. It is called machine-aided indexing, or MAI, and what it does, to put it very simply, is to compare the text of an abstract to a computer file of text strings called the Knowledge Base, which is a sort of enhanced Thesaurus. When it finds a match, it indexes the appropriate term, or terms, associated with the text string in the Knowledge Base.

Such a system is called a rule-based expert system, each element of text string with its indexing instructions in the Knowledge Base being referred to as a rule. What it produces is an index for editing on the text-editing system. Though still in the developmental stage, the system is already being used to index the abstracts appearing in three of the literature bulletins at CAIS.

As the MAI system's performance gradually improves through enhancement of the Knowledge Base, less and less material will be indexed by a human indexer. Obviously, this will save considerably on the cost of indexing. However, while the indexing done by the MAI system is editable and very consistent, its overall quality is not expected to approach that produced by a human indexer for quite some time.

This means that for the foreseeable future a part of the savings in indexing will go into increased editing time. And that means that additional members of the indexing staff will have editorial responsibility. In fact a new breed of indexer/editor will result who will be editing the output of a machine, on the one hand, and contributing to the growth of the Knowledge Base, on the other.

SPECIAL PAPER

# Wheeler's Gift of Electrical Books at the Engineering Societies Library: A Legacy and a Responsibility

Ronald R. Kline
Joyce E. Bedi
Thomas D. Lindblom

**SUMMARY.** The Wheeler Gift of books, pamphlets, and periodicals in the Engineering Societies Library in New York City was surveyed in 1985. Spanning four centuries of electrical history with over 6,000 titles, the Wheeler Gift was assembled in the late 19th-century by Josiah Latimer Clark, an English telegrapher and bibliophile, and was purchased and donated to the American Institute of Electrical Engineers by Schuyler Skaats Wheeler in 1901. This report gives a history of the collection, which is now dispersed throughout the Library, presents the results of the survey, and makes

Ronald R. Kline is Director of the IEEE Center for the History of Electrical Engineering, 345 E. 47th St., New York, NY 10017. He received a PhD in History of Science from the University of Wisconsin.

Joyce E. Bedi is Curator of the Center for the History of Electrical Engineering. She received a MA in Material Culture from James Cook University of North Queensland, Townsville, Qld., Australia.

Thomas D. Lindblom was Research Assistant at the Center for the History of Electrical Engineering in 1985. He received a BA in History of Science and Technology from Hampshire College.

 

several recommendations for the preservation of Clark's remarkable library.

The recent growth in the field of electrical engineering history has renewed interest in the Wheeler Gift of books, pamphlets, and periodicals, one of the most important library collections on the subject. Spanning four centuries of electrical history with over 6,000 titles, the Wheeler Gift was presented to the American Institute of Electrical Engineers (AIEE) in 1901 and is now housed in the Engineering Societies Library (ESL) at the United Engineering Center in New York City. Although about 800 items are in an area of locked shelving, the bulk of the collection was shelved by subject throughout the ESL over 70 years ago. Since it was unknown how many items had been lost or had deteriorated through normal use, a survey was undertaken in the summer of 1985 to locate and assess the condition of each item listed in the Wheeler catalog. This paper provides a history of the Wheeler Gift, followed by details of the survey and its results.

## *HISTORY OF THE WHEELER GIFT*

The collection of books and pamphlets that became known as the "Wheeler Gift" was assembled by Josiah Latimer Clark (1822-1898), a noted English telegrapher and bibliophile. His high standing in electrical circles was recognized in 1875, when he was elected the fourth president of the Society of Telegraph Engineers, the forerunner to the present Institution of Electrical Engineers (IEE).[1] Clark endeavored to compile a complete library of electrical literature published in England, adding any titles he could obtain from the Continent and America. After nearly half a century, he had assembled over 6,000 titles in six languages. In 1901, three years after Clark's death, the availability of his library became known to American engineers. The *Electrical World* "hoped that some public-spirited American citizen will try and obtain it for this country, and not let it go, as it easily might, into the hands of Japanese and German representatives of electrical associations, who have, it appears, cast upon it longing glances for a long time past."[2]

The "public-spirited" American who came forth was Schuyler Skaats Wheeler, a successful electrical manufacturer. Wheeler's interest in the Clark library dated back to 1885, when he and

T. Commerford Martin, editor of *Electrical World*, discussed the "Latimer Clark collection as the element and foundation" with which to establish a library for the AIEE.[3] The plans lay dormant until the collection came on the market in 1901. Then Andrew Carnegie and Wheeler were both approached to help purchase Clark's books. Wheeler, anxious not to lose the collection to Japan, bought the library in February 1901 for $6,880.28 and donated it to the AIEE in May. Carnegie matched this sum to cover maintenance and cataloging expenses.[4]

Wheeler conditioned the gift on the Institute's acceptance of five provisions:

1. To insure the library against fire and provide $1,500 annually for the maintenance of the collection.
2. To complete a *catalogue raissonne* explaining the features of interest of each book, and to send a copy of the catalog to each AIEE member.
3. To appoint a library committee of AIEE members.
4. To keep the library in New York City and available to all.
5. To exhibit rare books under glass and allow them to be used only under a librarian's supervision.

Wheeler's final request, an unofficial sixth deed of gift, was "that the Institute shall within five years, raise a sufficient fund by subscription, and provide itself with a permanent home for its meetings and library, and that this home shall be centrally located, reasonably safe from fire and not heavily mortgaged." [5]

The conditions of Wheeler's Gift figured prominently in the subsequent negotiations leading to the establishment of an American engineering center and library. The talks began at the AIEE's annual dinner in February 1903, and, less than a week later, Carnegie penned a brief note allotting a million dollars, "more or less," to construct one building for the AIEE, American Society of Mechanical Engineers (ASME), American Institute of Mining Engineers (AIME), and the American Society of Civil Engineers (ASCE). After the "civils" decided not to take part in the project, most likely because they had built a new headquarters building just seven years earlier, Carnegie gave one and one-half million dollars, on March 14, 1904, to build a "Union Home" for the AIEE, ASME, AIME, and the Engineers' Club.

On May 11, 1904, the United Engineering Society (UES), the

forerunner of the present United Engineering Trustees, was incorporated with the object to advance the "engineering arts and sciences in all their branches and to maintain a free public engineering library."[6] Shortly thereafter, the UES purchased property at West 39th Street in New York City, and arranged for the design and construction of a joint engineering building. The cornerstone was laid on May 8, 1906, and the 13-story Engineering Societies Building was ready for occupancy by January 1, 1907.

Early that year, the libraries of the three Founder Societies (ASME, AIEE, and AIME) were moved to the top two floors of the new building, where they formed the Library of the Engineering Societies. It was far from a joint library, however, because each society's collection was kept in its own section of the library on the twelfth floor and had its own staff and cataloging system. The main portion of the AIEE's collection was the Wheeler Gift, housed from 1901 to 1906 at the Institute's offices at 195 Liberty Street. In 1908, as the cataloging of the Gift was being completed, Wheeler suggested that the "Latimer Clark Library" be placed where it would be "more accessible and more easily seen than at present, preferably in the main library room on the thirteenth floor," which was also the reading room.[7]

This apparently did not occur, and the Library eventually grew into the type of cooperative venture envisioned by Scott and others. Although the UES hired a chief librarian in 1908, the Library still operated under the old confederation of society libraries for several years. Then, in 1913, the UES revised its bylaws, placing the Library under the control of a Library Board, instead of the library committees of each society. This led to a Library Agreement between the UES and the three Founder Societies in 1915, which was revised in 1916 to accommodate the addition of the ASCE's library. An extensive project to recatalog the entire collection under one system was begun in 1919. The project took seven years, during which time the Wheeler Gift was removed from the AIEE stacks in the Library and merged with the rest of the holdings according to subject heading.[8] Sometime after the 1930s, rarest and most valuable Wheeler books, approximately 800, were placed under lock and key.

The AIEE Library Committee allotted one-half of Carnegie's $7,000 matching grant to catalog the Wheeler Gift and chose the Reverend Michael Francis O'Reilly, a Christian Brother and physics professor at Manhattan College, for the job. O'Reilly, known as

Brother Potamian in the Order,[9] annotated each title in the collection, expanding on the three-volume, handwritten catalog kept by Clark and his librarian, a task that took several years. In 1906, Wheeler, as president of the AIEE, appointed William D. Weaver as a "Committee of One" to oversee the completion of the project.[10] The resulting two-volume *Catalogue of the Wheeler Gift of Books, Pamphlets and Periodicals in the Library of the American Institute of Electrical Engineers,* published in 1909 and edited by Weaver, fulfilled the library committee's expectations for a work that "promises through its annotations to have a unique value aside from its character as a catalogue."[11] For anyone interested in early electrical literature, Brother Potamian's description of the Clark Library remains an indispensable guide.

The two-volume catalog is divided into eleven sections, each of which is arranged chronologically. Section I, "Main Portion of Collection," fills the first volume and lists over 2,700 titles, dating from 1473 (*Speculum Naturale* by Bellovacensis Vincentius) to 1906. In addition to his detailed annotations of titles up to the 18th century, Potamian reproduced several frontispieces and illustrations from the books to enrich the catalog. Section II, "Excerpts from Periodicals—Miscellanea," lists just under 2,000 reprints from scientific and technical journals beginning in 1704. The next eight sections of the catalog total over 1,500 near-print entries divided between:

- III. Instructions, Rules, and Regulations for Telegraph Operation—Tariffs—Codes
- IV. Reports of Telegraph and Cable Companies
- V. Prospectuses of Telegraph and Cable Companies
- VI. Reports of Electric Light, Telephone and Manufacturing Companies
- VII. Patent Specifications—Litigation
- VIII. Parliamentary Papers—Legislation—Legal
- IX. Expositions—Congresses—Societies—Banquets, etc.
- X. Trade Catalogues, Circulars and Price Lists

Section XI of the *Catalogue* lists over 120 periodicals arranged alphabetically. For over 75 years, Potamian's *Catalogue of the Wheeler Gift* has served as the only comprehensive description of the collection.

## *THE SURVEY*

### *Methodology*

The primary goal of the survey was to locate, if possible, each item listed in Potamian's *Catalogue*. In order to do this, the call number of each Wheeler entry was determined from the ESL card catalog, which is based on an expanded version of the Brussels, or Universal Decimal Classification, system created especially for technical and scientific literature. The call number was then used to find the item in the ESL stacks. Particular attention was given in the course of the survey to locating missing titles, i.e., those that were either not in the ESL card catalog or were not found on the shelf. One-third to one-half of the items first considered missing, mostly pamphlets, were found by the end of the survey. This left a total of 532 missing Wheeler entries.

It was also decided to include an indication of the overall condition of the Wheeler Gift as part of the survey. To this end, each item was examined and given a rating of "good," "fair," or "poor," based on the simple criterion of how well a volume could withstand further use. Basically, a "good" book shows little deterioration; its binding and pages are intact and supple, there is no sign of insect or water damage, etc. Many of these "good" books have been rebound, which has afforded them additional protection over the years. Typically, a "fair" book is somewhat brittle, with taped or tattered bindings, but, when handled with care, will not significantly deteriorate with continued use. A "poor" rating was reserved for a book that will be damaged by any further use; brittle pages disintegrate at the touch, the binding or cover is detached, the volume is held together with string, the leather binding is crumbling from "red rot," etc.

During the course of locating and assessing the Wheeler entries, any additional important information, such as references to other works or details of physical condition, was also noted. All of the information on the books found during the survey was stored in a computer database from which several statistics have been compiled.

### *Results*

Figures 1, 2, and 3 and two appendices detail the results of the survey. The three figures describe the Wheeler Gift by year,

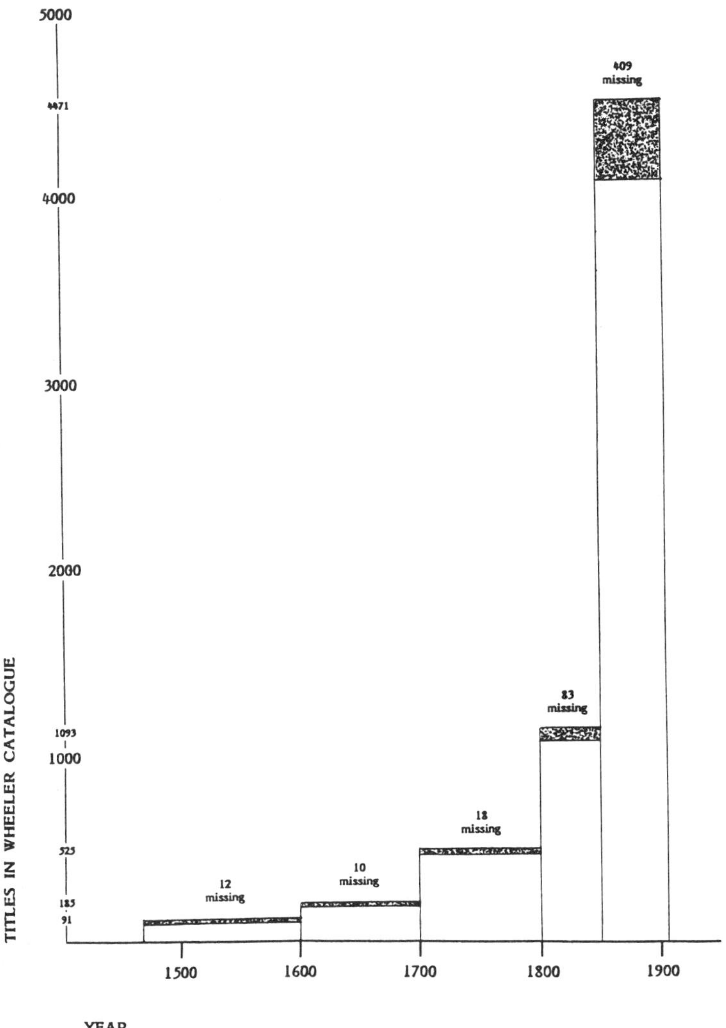


Figure 1. Wheeler Gift by Year

condition, and subject. Appendix A lists the volumes which could not be located, while Appendix B, arranged by Potamian's *Catalogue* entry number, lists the ESL call number, condition of the book, and comments. Copies of Appendices A and B have been deposited with the Engineering Societies Library and the Center for the History of Electrical Engineering.

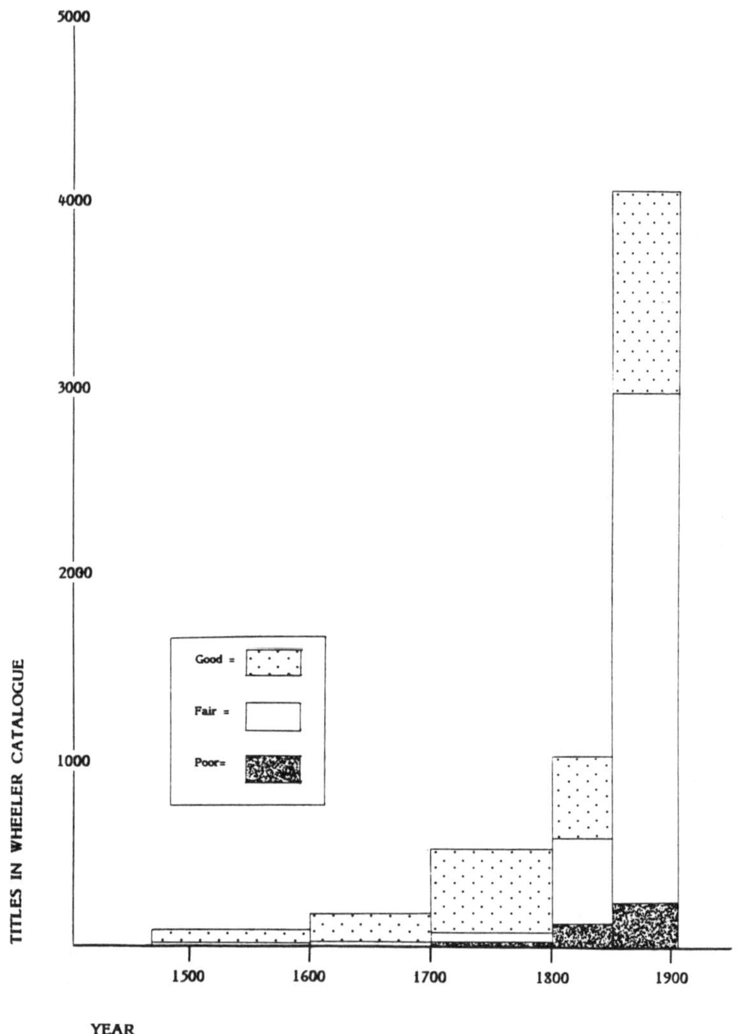


FIGURE 2. Wheeler Gift by Condition and Year

A word should be said about the "Year" axes of Figures 1 and 2. The bar graphs represent the number of entries listed in the *Catalogue* for five time periods. However, the *Catalogue* does not list entries in exact chronological order by year of publication. In a few cases, Potamian grouped all items related to one author under the first publication by that author. For example, the holdings

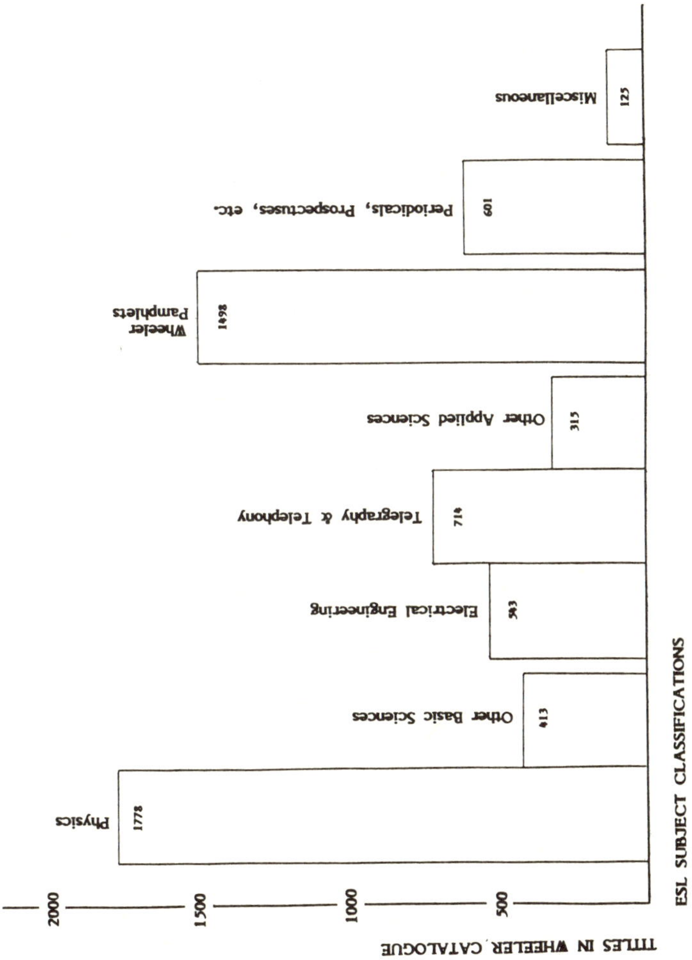

TITLES IN WHEELER CATALOGUE

2000
1500
1000
500

Physics 1778
Other Basic Sciences 413
Electrical Engineering 543
Telegraphy & Telephony 714
Other Applied Sciences 315
Wheeler Pamphlets 1498
Periodicals, Prospectuses, etc. 601
Miscellaneous 125

ESL SUBJECT CLASSIFICATIONS

FIGURE 3. Wheeler Gift by Subject

relating to Gilbert are listed sequentially from the year 1600 onward in the 17th-century portion of the *Catalogue*, even though several of these entries are biographies or translations of Gilbert that were published in the late-19th or early-20th centuries. Thus, since Potamian's listing was used in manipulating the data, the numbers do not exactly reflect the books published in a given time. The difference, however, is small and is almost nil after 1800.

### *Missing Wheeler Entries*

Figure 1 outlines the Wheeler Gift by year, and includes totals for the missing items. Less than 9 percent of the collection (532 items) is missing. Due to Potamian's system of arrangement of the *Catalogue*, as mentioned above, only 31 of the 40 missing volumes noted in the first three columns of the graph were actually published before 1800. The majority of the other missing entries, particularly the 409 items from the period 1850 to 1906, are reprints and pamphlets.

### *Condition Assessment*

Figure 2 addresses the artifactual value of thc Wheeler Gift by summarizing the condition ratings designated during the survey. The majority of the Wheeler entries published before 1800 are in good condition, since the ESL has kept approximately 800 old and rare books from the Wheeler Gift separately in locked shelving for many years and also had many of the volumes rebound, by a custom bookbinder, in the 1930s. In addition, the quality of the paper used in books printed prior to 1850 is usually better, and it can be assumed that these books received more attention from Clark himself as already being significant historically.

The fourth column of the graph, which shows the increase in electrical literature after the discovery of a constant source of current in 1800, shows slightly fewer books in the "good" category, perhaps due to a combination of poorer original quality, greater use over time, and storage conditions. The bulk of the Wheeler Gift (roughly 4,200 titles, reflected in the fourth and fifth bars of Figure 2) resides in the regular ESL stacks. The system of arranging these shelves by subject means that a book from the Gift can be placed next to a heavily-used modern volume. The abrasion and stress placed on the older binding, caused by pulling and reshelving the newer book, can hasten the deterioration of the Wheeler volume without its ever being used.

The last bar of the graph dramatically points out that the majority of the Wheeler entries, over 4,000, were published after 1849. This period includes all of the electrical reprints compiled by Clark from the *Philosophical Magazine* (4 volumes) and from the *Philosophical Transactions of the Royal Society* (15 volumes) which, together, number almost 600 titles. It also includes most of the 1,500 "Wheeler Pamphlets" (an ESL designation for the majority of the near-print material). These publications were rated generally as "fair," but definitely at the lower end of that designation. While the paper in these pamphlets and reprints has held up surprisingly well, the bindings have seriously deteriorated. In the majority of cases, the volumes are only being kept intact by the string with which they are trussed up.

The "Wheeler Pamphlets" deserve a special note. This portion of Clark's collection originally consisted of 3,450 titles, which were bound into 195 volumes, each with a table of contents. Well before the ESL moved to its present location, over 2,000 of these pamphlets were removed from the original bindings, rebound into sturdier volumes, and assigned ESL call numbers and titles such as "Electric Currents, Induction, Etc.—Reprints." The remaining 1,500 pamphlets are the most deteriorated group of material from the Gift. Their original bindings are simply wrapped around them like folders, with string tied around the bindings. The bindings thus do not provide much physical support or protection, and the edges of the pamphlets are frayed and broken as they scrape against the string. This near-print material is nearly impossible to replace. To its credit, the ESL recognized both the value and the sad condition of the pamphlets, and first moved the collection into the area of locked shelving. A project to clean, wrap, and box the volumes, using acid-free materials, is in progress; the first 60 of the 195 volumes have been completed.

The condition of the "Wheeler Pamphlets" points out preservation problems common to all of the holdings of the ESL. The entire collection suffers from a scarcity of environmental controls. Temperature and humidity fluctuations in the stacks are great, since the central heating/air conditioning in the United Engineering Center, home of the ESL, is shut off at night and on weekends. The only humidity control and air filtration is that inherent in the central system. This combination of ever-changing temperature and humidity levels and of airborne pollutants endemic to the New York City area accelerates the deterioration of bindings and of text areas exposed to the air, and places much physical stress on the books. In the event of fire, the collection is protected by an overhead sprinkler system. However, the water pipes running above the stacks have leaked in the past. Poor housecleaning in the stacks leaves dirt on the books which, in turn, provides anchors for moisture and microorganisms. Working with such handicaps, the ESL staff is to be commended for the overall satisfactory condition of the Wheeler Gift.

### *Subjects*

Figure 3 describes the collection according to subject, as represented by ESL call numbers. The most prominent column, ''Physics,'' depicts entries related mainly to electricity and magnetism (call numbers 537 and 538). However, the best represented subject in the collection is British telegraphy. Brother Potamian compiled an ''Index to Telegraphic Entries'' for the *Catalogue*, in which he lists over 2,000 titles. These appear in several columns in Figure 3. Most of the 700 items cataloged under the call number 621.38 (''Telegraphy & Telephony'') and approximately 600 ''Wheeler Pamphlets'' cover telegraphy. Additionally, three volumes of ''Prospectuses of Telegraph Companies'' total over 200 titles. The remaining 500 telegraphic entries are distributed under call numbers relating to ''Electrical Engineering'' and ''Physics.'' The other subject headings included in the graph also reflect Clark's interests:

Electrical Engineering other than telegraphy and telephony (electrometallurgy, lighting, batteries, etc.)
Other Basic Sciences (natural philosophy, chemistry, mathematics, etc.)
Other Applied Sciences (mechanical and civil engineering)
Miscellaneous titles outside the physical sciences (geography, cartography, and history).

The content of the books and pamphlets in the collection offers the historian of electroscience and technology a wealth of information in all areas of the subject up to the late 1880s. As mentioned above, the greatest emphasis of the Gift, due to Clark's professional interests, lies in electric telegraphy. The earliest theoretical and practical work of Soemmering, Ampere, and Morse is well represented. The legal disputes between Cooke and Wheatstone and the voluminous legislation from the British government regarding telegraphic regulation and ownership abound in the collection (see Section VII of the *Catalogue*). Prospectuses and reports to shareholders of both land and submarine telegraph companies, cataloged mainly in Sections IV and V, number over 300 titles.

Clark's collection has over 500 titles in the other areas of 19th-century electrotechnology, including lightning rods and electrometallurgy, electric batteries, lighting and power, and the telephone. He also collected documents relating to the economic development of these technologies. For example, Section VI of the *Catalogue*, ''Reports of Electric Light, Telephone, and Manufacturing Companies,'' contains 157 references to British firms.

Historians interested in the exchange of knowledge between science and technology can benefit from the Wheeler Gift's exceptional strength in the earliest theoreticians and experimenters in electrophysics—beginning with the treatises of Pliny, della Porta, and Gilbert, through Descartes, and into the 18th-century holdings of Swedenborg, Musschenbroek, Nollet, Franklin, and Coulomb. The invention of the voltaic pile in the beginning of the 1800s opened the new fields of constant current electricity and electromagnetism. This period of growth is represented in the Wheeler Gift by such standard treatises as those by Volta, Oersted, Ampere, Faraday, and Maxwell, as well as the works of hundreds of lesser known physicists. Clark also collected the work of his contemporaries who advanced the science of electricity, among whom are Silvanus Thompson, the Siemens brothers, Lord Kelvin, Wheatstone, Helmholtz, Gherardi, and Becquerel.

Brother Potamian made annotations for all the reprints, prospectuses, company reports, and the other near-print items now called the ''Wheeler Pamphlets.'' His notes serve as the best source of information about these rare and important documents. Several series of rebound ''Wheeler Pamphlets,'' with broad titles like ''Magnetism and Electricity Reprints,'' ''Ayrton and Perry Papers,'' and ''Lightning and Lightning Conductors—Angelelli and Others,'' represent the range of subjects, languages (Latin, French, Italian, German, and English), and levels of treatment (from the scientific to the economic to the legislative) available in the collection. This material also presents a strong cross-section of Clark's collection. Historians interested in early English telegraphy, the development and acceptance in the international scientific community of the concept of electromagnetic induction, or a number of other topics in the history of electricity can find substantial source material in the "Wheeler Pamphlets."

The 128 periodicals listed in the final section of the *Catalogue* comprise a major resource of primary published material in all areas of electrical science and technology. Besides the complete runs of such standard journals as *Philosophical Transactions of the Royal Society* and *Scientific American*, the collection contains more uncommon telegraphic and electrical journals in English, French, Italian, and German, including *American Electrician* (1896-1905), *Bulletin International de l'Electricite* (1882-1895), *Italy—Ministero Delle Poste e dei Telegrafi—Relazione Statistica* (1891-1901), and *Zeitschrift des Deutsch-Oesterreichischen Telegraphen-Vereins* (1854-1869).

## COMPARISON TO OTHER COLLECTIONS

The Wheeler Gift compares very favorably with five other major collections of electrical literature, three of which are in the United States, while two are in England. The first of these was given to the Massachusetts Institute of Technology in 1912 by Theodore Newton Vail, President of the American Telephone & Telegraph Company, in the name of that firm.[12] The portion of the gift now known as the "Vail Rare Collection" contains 3,050 titles published from the 16th to the early 20th centuries, a significant number of which cover electricity and magnetism. The "Vail Pre-1910 Collection" contains 12,350 titles, approximately 4,000 of which cover electrical engineering. In addition, a large number of pamphlets relating to electrical matters and about 500 titles on animal magnetism are in the "Vail Rare Collection." The remainder of the original gift, which also contains much electrical engineering material, formed the nucleus of the Barker Engineering Library and still circulates.

The Bakken Library of Electricity in Life, in Minneapolis, MN, maintains a book, journal, and manuscript collection of some 12,000 items focusing on the historical role of electricity and magnetism in the life sciences and medicine. Included are works by Nollett, Franklin, Galvani, Aldini, Volta, and others. The Bakken's holdings of ephemera—advertisements, circulars, and instructional pamphlets from the 19th century—number approximately 200. The library also retains nearly 300 trade catalogs and price lists printed between the years 1875 and 1930, as well as a versatile collection of secondary histories, biographies, and reference works.

The Bern Dibner Collection, the third comparable assemblage of electrical books, is divided between the Burndy Library in Norwalk, CT, and the Smithsonian Institution Libraries in Washington, DC. The Burndy Library has approximately 7,500 titles covering electrical science from the earliest times to the present, many of which are pamphlets and reprints. The remainder of the collection, nearly 11,000 pre-1870 titles, was given to the Smithsonian in 1976; approximately 3,000 of those titles cover electrical topics.[13]

The library most similar in scope to the Wheeler Gift, however, is the Sir Francis Ronalds collection at the Institution of Electrical Engineers in London. Knighted in 1870 for his early contributions to telegraphy, Ronalds was also an ardent bibliophile. He devoted the last 20 years of his life to book collecting in England and Italy, seeking to compile a complete electrical library. He died in 1873, at the age of 85, having amassed a collection of 2,000 books and 4,000 pamphlets.[14]

A comparison of the catalogs of the two libraries shows that Clark's collection is stronger in treatises on the earliest electrical subjects—works on the lodestone, the mariner's compass, and terrestrial magnetism. Clark also purchased more books by natural philosophers up through the 17th century, but Ronalds's holdings of 18th- and early 19th-century continental titles exceeds those of Clark. Otherwise, the science of static electricity, up to Volta's invention of the battery, is evenly represented in the two collections, as is that of dynamic electricity up to the 1860s. But because Clark collected electrical books for another 20 years after Ronalds died, the Wheeler Gift far exceeds Ronalds's collection in literature from the latter part of the 19th century, when numerous practical applications of electricity were made. This is particularly true in regard to electrical periodicals, whose number increased substantially after Ronalds stopped collecting. The fact that the ESL has kept many of these journals up to date enhances their research value.

Also housed at the IEE library in London is the collection of Silvanus P. Thompson (1851-1916), a prominent electrical engineering educator in Britain. Thompson's library includes about 900 books published prior to 1825, more than one-quarter of which date from the 15th to the 17th centuries. Many are "association" copies of historical interest, for example, Volta's copy of Galvani's *De viribus electricitatis*, given to Volta by the author. The remainder of the collection contains about 2,300 books published after 1825. The bulk of these (1,407) are from the late-19th and early-20th centuries (1880-1909). They deal mostly with electricity and magnetism, but also cover the topics of optics, acoustics, heat, mathematics, astronomy, engineering, and chemistry. Several periodicals are included, as well as 8,000 pamphlets, comparable in number to those collected by Clark.[15]

## *CONCLUSION*

This survey has provided both gratifying and disturbing results. It is gratifying to know that the Wheeler Gift is still very much intact within the ESL; approximately 90% of the collection was located. It is also gratifying to know that the rarest books are still in good condition, and that the ESL has done its best to ensure the safety of the volumes. It is disturbing, however, to learn that several titles of both historical and financial value are missing. It is also disturbing to realize that the majority of the Wheeler Gift is in fair to poor condition. Despite preservation efforts undertaken by the ESL staff, the collection continues to deteriorate.

These results, however, can be used to make some recommendations for the future of Clark's remarkable record of electrical science and technology.

1. The preservation work carried out by the ESL staff is to be commended and encouraged. It is suggested that this work, mainly the construction of protective enclosures, continue on a regular basis, and that priority in this work be given to those items identified during the survey as being in "poor" condition. To ease the strain on the already over-extended library staff, the possibility of a cooperative arrangement with the Center for the History of Electrical Engineering in performing this work is also suggested.
2. The "Wheeler Pamphlets" could be dealt with as a special project for which grant support could be sought. It is recommended that such a project include the cleaning, wrapping, and boxing of the remaining 135 volumes. It would also be useful to make this unique material more accessible and better known by updating the five card catalog drawers

pertaining to the pamphlets, searching for the missing items, and preparing an index to the "Wheeler Pamphlets."

3. It is also recommended that the ESL increase its vigilance in regard to basic housecleaning in the stacks. Air filters need to be changed or cleaned regularly, and stack areas require a thorough dusting and vacuuming. Maintaining a schedule of good housekeeping is still one of the most important facets of collection care.
4. Finally, it is recommended that consideration be given to reassembling the Wheeler Gift. The collection is one of the real gems of the ESL, and having it together would both enhance its worth, especially by increasing its security and preservation, and raise its visibility. Ideally, the collection should be housed on baked-enamel finish, metal shelving, in a separate, secure room, with provisions for constant temperature and humidity control, air filtration, ultraviolet-shielded lighting, fire protection (preferably a Halon system), environmental monitoring (recording thermometer/hygrometer), and work space.

Latimer Clark knew the value of his library, and, as a trustee of the Ronalds Library at the IEE, thought his own collection would be of more use to researchers outside of Great Britain. Through the generosity of Schuyler Skaats Wheeler, that role was assured. The legacy of the Wheeler Gift, however, is also a responsibility to care for the collection, making sure that it is available for researchers in years to come. It is hoped that this survey will help fulfill that obligation.

## REFERENCES

1. Finn, Bernard S. Josiah Latimer Clark. *Dictionary of Scientific Biography*. New York: Scribner's, 1971, vol. 3, pp. 288-289.

2. Latimer Clark Library. *Electrical World*, 37: 300, 1901 February 23.

3. Library Dinner. *Transactions of the American Institute of Electrical Engineers*, 21: 97-128, 1903 February 9 (p. 109).

4. O'Reilly, Michael Francis (Brother Potamian). *Catalogue of the Wheeler Gift of books, pamphlets and periodicals in the Library of the American Institute of Electrical Engineers*, ed. William D. Weaver. 2 vols. New York: AIEE, 1909, vol. 2, pp. 468-469.

5. Wheeler, Schuyler Skaats. Deed of Gift. 1901 May 17: p. 4. Copy in IEEE Archives, Center for the History of Electrical Engineering, New York, NY.

6. Mount, Ellis. *Ahead of its time: the Engineering Societies Library, 1913-1980*. Hamden, CT: Linnet Books, 1982, p. 45.

7. Stott, H. G. (Letter to S. S. Wheeler). 1908 June 3. Copy in IEEE Archives, Center for the History of Electrical Engineering, New York, NY.

8. Mount, pp. 61-75, 129-134.

9. Curran, Edward J., Brother Potamian: Rev. Michael Francis O'Reilly. *Records of the American Catholic Historical Society of Philadelphia*, 28: 202-213, 1917.

10. Wheeler, Schuyler Skaats (Letter to Ralph W. Pope). 1906 August 16. Located in IEEE Archives, Center for the History of Electrical Engineering, New York, NY.

11. O'Reilly; vol. 2, p. 469.

12. Lane, Ruth M. The Vail Library at Technology. *The Tech Engineering News*, 6: 201, 236, 238, 240, 242, 244, 246, 1923 December.

13. Smithsonian Institution Libraries. The Dibner Collection: Main Entry List. 1979. Unpublished report.

14. Symons, E. D. P. Two electrical book collections: The Ronalds and Thompson Libraries. *Institution of Electrical Engineers Proceedings*, 132(A): 582-586, 1985 December.

15. Symons, pp. 583-586.

## ACKNOWLEDGMENTS

In conducting the survey and preparing this report, the authors would like to acknowledge the cooperation, assistance, and advice of Kirk Cabeen and Ari Cohen, Director and Head Cataloger of the ESL, and the general guidance of the IEEE History Committee, chaired by Harold Chestnut. Information on similiar collections was supplied by Elizabeth Ihrig, Librarian, Bakken Library of Electricity in Life, Minneapolis, MN; Kathy Marquis, Reference Archivist, MIT Libraries, Cambridge, MA; E.D.P Symons, Archivist, Institution of Electrical Engineers, London, England; Philip J. Weimerskirch, Assistant Director, Burndy Library, Norwalk, CT; and Ellen Wells, Librarian, Dibner Collection, Smithsonian Institution Libraries, Washington, DC.

A special note of thanks goes to the IEEE Life Members, who, through the Life Member Fund Committee, chaired by Julian Tebo, funded the summer intern for 1985, Thomas Lindblom, who carried out the survey of the Wheeler Gift.

# Preserving the Collections of the Science and Technology Research Center of the New York Public Library

John P. Baker

**SUMMARY.** The collections of the Science and Technology Research Center, a unit of The New York Public Library's Research Libraries, are described. The collections are estimated to contain 1.2 million volumes. Causes of deterioration and the administrative context of preservation planning are presented. Past and current efforts to preserve the collections, which are composed largely of little-used publications, are delineated. Planning for the future includes reintegration of the collections which are now housed in two separate locations at a major new facility to be located at the Library's West 43rd Street Annex, initiation of a systematic collections cleaning, refurbishing and repair program, and a stepped-up preservation microfilming program in cooperation with other research libraries in the nation.

## *INTRODUCTION*

A few months before the celebration marking the Fiftieth Anniversary of the opening of the Central Building of The New York Public Library, curators of several of the major research divisions were asked to give their views concerning the various types of user needs that are served by the Library. Their observations were included in a publication that appeared later the same

---

John P. Baker received the BA degree from Boston University in 1955 and the MLS from Columbia University's School of Library Service in 1965. Since that time, he has worked in The Research Libraries of The New York Public Library, first in public service divisions as a reference librarian, and since 1968, in administrative positions. In 1972, he was appointed Chief of the newly-created Conservation Division, and in 1985 was appointed Assistant Director for Preservation. His responsibilities include management of all preservation programs within The Research Libraries including preservation microrecording, binding, shelf preparation, collection maintenance, and conservation/restoration.

 

year in a volume entitled *Search and Research; the Collections and Uses of The New York Public Library* (1961). In the course of his remarks, George S. Bonn, Chief of the Science and Technology Division, observed, "everything at the Library is collected with the idea that it will become historical. The astronautical material that we are getting now will be historical in another ten years."

Bonn's view of the research library as a storehouse of historical materials is particularly relevant in considering collections of scientific and technological materials. Although scientific investigation necessarily mandates access to a large body of current information, most large research libraries that are collecting in these fields today accept the responsibility as well for maintaining strong retrospective collections, to meet the unpredictable needs of public inquiry. The Mission Statement of NYPL's Research Libraries takes a yet more comprehensive view by stating that The Research Libraries are to serve as a "library of record; to use its available resources in a balanced program of collecting, organizing, and conserving books and other materials, and to provide ready access to library users in New York City, and through cooperating libraries and library networks, to users throughout the country and the world."

The Science & Technology Division, renamed the Science and Technology Research Center (STRC) in 1972, is part of a larger family of reference and research facilities located in Manhattan that is known collectively as The Research Libraries. For the most part privately funded, The Research Libraries' major sources of support are endowed funds provided by the Astor, Lenox, and Tilden Foundations of The New York Public Library and annual gifts received from corporations, private foundations and individuals. Over $5 million annually is provided by New York State, and this is supplemented by grants in varying amounts received from the two National Foundations for the Arts and the Humanities, the U.S. Office of Education and other state and federal government granting agencies. The City of New York, which owns and maintains the Central Building, provides funds for its maintenance and security, for major capital improvements and—beginning in 1986—money to help preserve materials related to the history of New York City.

This article focuses on one aspect of NYPL's mission: the task of conserving (or preserving—the two terms are virtually synonymous) the collections of The Science & Technology Research Center.

Simply stated, the preservation problem faced by research libraries having huge retrospective collections derives from the fact that paper, microforms and most other information-bearing media they collect are composed of organic materials that are subject to gradual deterioration and—if not properly cared for—to eventual total disappearance by crumbling to dust. The poor quality of paper produced for publishing purposes in the last 150 years, inadequate binding methods and materials and the hazards imposed on fragile library materials by careless use and unsatisfactory storage conditions are in league with the natural tendency of organic materials to self-destruct over time. This perilous combination of forces can be effectively thwarted by interposing various preservation measures that will retard the deterioration process indefinitely. The process cannot be halted completely, however, since most materials must be kept available for use. The only guaranteed method of stopping deterioration completely is to store objects in darkness, at low temperature, in an oxygen-free, inert gaseous atmosphere such as argon.

Preservation is the essential element in maintaining collections of historical materials. It is the over-arching and indispensable bridge, linking the past to the present, and both to the future. Preservation is the enabling instrument that assures that the student, the scholar and the independent investigator will have access to humanity's intellectual records in the years to come.

Great research libraries will continue to maintain large collections of retrospective materials, even though much of the material they retain is subject to declining use. From the library point of view the obsolescence factor, as it is sometimes called, is exacerbated by the rate of growth of scientific literature. In some fields the amount of information published has probably at least doubled in the last twenty years. Moreover, publications in various fields become obsolete at various rates. In a study published in 1970 in *Bioscience*, it was found that the highest possibility that a paper will be cited in another paper occurs within three years of its publication. After the first three years the probability of a citation declines by half in about fifteen years. Nevertheless, as Robert G. Krupp, one of the former Chiefs of the STRC put it recently, "use is not a function of value. The collections must be kept available for that *one* use which may be requested."

The economic challenge implicit in maintaining and preserving vast hordes of little-used materials is being addressed in some degree, and must be more actively addressed in the future, by relying heavily on such strategies as preservation microfilming, remote site compact storage at low temperature, fast and efficient interlibrary loan service, systems of rapid facsimile transmission and cooperative agreements among a number of institutions for apportioning equitably the burden of retention responsibilities. The newer reduced-image technologies such as analog disc must also be included in planning for the future.

## *OVERVIEW OF THE COLLECTIONS*

The Science and Technology Research Center houses one of the world's great research collections in the pure and applied sciences and related technologies: more than 1.2 million volumes and 4,800 current serial titles used by 80,000 people annually. The collections are housed in two locations, the famous Central building of gleaming marble that stands at the corner of Fifth Avenue and 42nd Street, and the Annex Building at 521 West 43rd Street, between 10th and 11th Avenues. The entire Patents Collection, thousands of volumes of science-related materials such as publications of learned societies, and the entire collection of some 80,000 volumes related to cookery and food technology are housed at the Annex. Budget and space constraints over the past fifteen years have resulted in collection fragmentation, as more and more material was relocated. At time of writing, it is estimated that 60 to 70 percent of the collection is in the Annex.

Approximately one quarter of the holdings represent pure and applied physical sciences, and about three quarters technology. The collections are rich in older as well as modern materials, and thus provide documentation for the history of subjects collected. In areas where current collecting policy is comprehensive, materials are collected from all countries and in all languages, including Oriental, Slavonic and Hebrew, although publications in non-roman alphabets are at present housed in the Oriental, Slavonic and Jewish Divisions, respectively.

The Patent Collection is second only in size to that of the United States Patent Office, and unlike it, includes strong foreign collections. STRC currently receives substantially complete collections of patents, including specifications and drawings from Belgium, Denmark, England, France, Germany and Sweden, in addition to those from the United States. In addition, abstracts of patents or patent lists are received from a number of foreign countries.

In a national context, the collections are considered exceptionally strong. Within the Research Libraries Group (RLG), a consortium of over 40 research libraries, STRC has primary collecting responsibility for no fewer than 20 out of 61 science and technology areas, including chemical technology, the tobacco, rubber, textile, leather, mining, and transportation industries—and of course, patents. This means that the Library has assumed the obligation among RLG members to maintain existing collections in these areas in usable condition and to continue its policy of comprehensive collecting in these subjects.

Even in areas in which STRC does not maintain significant collections—life sciences, biology and medicine—it still serves a highly useful service by maintaining basic collections in these areas, and by referring researchers to other collections in the New York area where their specialized needs can be met.

The need for better science education in this country has never been more apparent. At the same time, the nation is more and more dependent upon the availability and exchange of up-to-date information. This most contemporary mode of "commerce" is important to the economic health of the nation and essential to maintaining the vigorous intellectual health of its citizens. Since New York City is a center of undergraduate and graduate education, NYPL's Science and Technology Research Center is uniquely placed to become an even more important resource for science education and information retrieval in all its forms.

## *THE PLANNING ENVIRONMENT*

In considering the various factors that have an impact on preservation of the STRC collections, the following seem particularly deserving of comment. Some have a positive influence on preservation, others a less desirable one. Taken together they constitute the environment within which the preservation planning process must be carried on.

— Unlike many university libraries, the research collections of NYPL are a one copy library; when a copy is lost, mutilated, misshelved, or allowed to deteriorate beyond convenient use, and is neither replaced nor preserved, it is lost to the permanent collections.

— The research collections are the most heavily used in the world. Because of the Library's policy of providing free access to everyone who is 18 years or older, the collections have received heavy use by several generations of college and university students whose own institutional libraries are unable to provide the resources required for in-depth investigation.

— The collections are an important local and regional resource for business and industry. Many corporations maintain deposit accounts with the Library's Reprographic Services Division, which results in a substantial volume of face-down copying requests. As every librarian knows, this method of text reproduction is inherently damaging to library materials, though librarians tend to accommodate the demand as a necessary adjunct to information transfer. The problem is exacerbated at NYPL because such a large portion of the overall collection is on paper that has become brittle, and an exceedingly large portion of volumes in the collection have been bound using the oversewing method. This method produces a strongly bound book, but adds to the difficulty of obtaining face-down copies because much of the inner margin is lost during the binding process, with a resultant loss of flexibility.

— The Research Libraries have a closed stack policy, and with few exceptions do not participate in interlibrary loan.

— Although the shelving system that was adopted for permanent installation at the time of the Central Building's planning and construction during the period 1905-11 was a model for its day, it no longer serves adequately the needs of the collections. Oversized and odd-sized materials are poorly shelved, and bookends that were once an integral part of the original system design have long since disappeared from most areas of the stacks.

— Growth of the collections beyond the capacity of management to provide adequate space for housing them has resulted in the conservationally unsound practice of storing thousands of volumes on the fore edge; if volumes cannot be stored in an upright position, they should be stored on the spine as an alternative, although less desirable, practice. Volumes should never be stored on the fore edge, because gravity tends to pull the text block away from the spine to which it is secured.

— Until 1985 the central stacks were not served by an environmental control system, and the levels of temperature and humidity fluctuated widely according to the season; windows were kept open in warm weather to promote air circulation, thus permitting unfiltered air that was laden with gaseous pollutants of industrial and urban origin to circulate freely.

The preservation librarian must deal on a daily basis with the interplay of all of these factors and strive for improvement whenever and wherever possible. Given the intractable nature of some of these problems, there are rarely easy solutions that can be put into effect immediately.

## *ROLE OF THE CONSERVATION DIVISION*

Although the Conservation Division was not established as an administrative unit within The Research Libraries until 1972, the Library has a long history of leadership in the preservation of research materials. In the 1920s and '30s Harry Miller Lydenberg, Chief Reference Librarian, and later Director, participated in several national studies dealing with the causes of paper deterioration. His interest in the subject was shared by colleagues of his own generation who in turn passed their concern for preservation to their successors. In 1966 a Preservation Committee was formed as a standing committee of the Research Libraries Council, and later the same year the post of Collections Preservation Coordinator was created. In 1972 the Conservation Division was established, and was assigned responsibility for managing the first-time binding program, the preservation microfilming program and a fledgling collections maintenance and restoration program. Since 1972 the Conservation Division has had its share of ups and downs in marshalling support for effective preservation policies, improvements in the storage environment and provision of preservation and conservation treatment services commensurate to the needs of the collections and the fiscal realities of the Libraries' financial resources.

Today, well over $2 million is being spent annually on preservation activities, and the staff of the Conservation Division numbers

75 FTE positions. The Chief of the Conservation Division also serves as the Assistant Director for Preservation in The Research Libraries. He is responsible for the day-to-day administration of the Division, for reviewing existing policies and suggesting new policies and procedures that have a bearing on preservation of the collections, and for long-range planning. He plays a major role in identifying preservation needs throughout The Research Libraries and works closely with the Development Office and the Budget Office in pursuing funding for needs and priorities that have been identified by the administrative staff of the Library. In addition to other responsibilities, the Assistant Director for Preservation represents The Research Libraries in local, regional and national forums, and interprets the Libraries' preservation policies, goals and objectives to members of the staff and to the general public.

The Conservation Division includes major components for managing the first-time binding and shelf preparation program; collections maintenance and repair; preservation microfilming; a Photographic Laboratory; and a restoration unit. In addition to microfilming deteriorating publications, the Photo Lab produces photographs and transparencies in various formats on a public order basis and produces microfilm positive service copies of titles for which The Research Libraries own a master negative. The volume of public orders, received mainly from other libraries, has increased substantially over the past four years as The Research Libraries' holdings have become more widely accessible through records for master negatives that are being fed into the Research Libraries Information Network (RLIN). The Custom Binding and Restoration Office provides treatment for scarce and rare materials, objects selected for display in the Library's public exhibition program and materials from special collections including rare books, prints, posters, autographed music scores, manuscripts and photographs. Among the treatments it is able to provide at present are deacidification of individual sheets, polyester encapsulation, box-making, and book and flat paper restoration.

Judicious selection procedures and wise decisions as to the *form* of treatment that objects will receive are essential to the success of the program. Materials are selected by public service staff who have bibliographic expertise in various subject areas collected; the form of treatment is also decided by public service staff in consultation with various members of the Conservation Division.

The work of the Conservation Division ties in closely with the operations of many other units of the Library. Continuous contact is maintained with Division Chiefs, Curators and selection staff in twenty public service units; with the maintenance, security, personnel, budget, accounting and development offices of Central Services; and with the acquisitions, cataloging, collections management and development, public services and preparation services offices. Liaison with these units is supplemented by regularly scheduled meetings with the Committee on Conservation and Protection of the Collections of the Research Libraries Council. The Chief of the Division serves as Chair of the Committee and is also an ex-officio member of the Research Libraries Council, which meets once a month. Each of the public service units has a member of the staff who is designated Conservation Representative and maintains liaison as needed with the various administrative units of the Conservation Division. Two program meetings for Conservation Representatives are held during the year, at which special presentations on particular aspects of preservation may be made and discussion held on matters of common concern. A recent meeting included a report by a consultant who was asked to investigate policy and practice with respect to the production and storage of copy negatives generated over the years in response to public order requests for illustrative materials in custody of various public service units and the special collections.

Although much is being accomplished, the work of preservation is truly never-ending. The Andrew W. Mellon Preservation Administration Intern assigned to the Conservation Division Administrative Office has commenced investigations leading to development of a new five-year plan designed to bring the existing program within sight of realizing nationally-recognized preservation norms and objectives. A major project for collections analysis and improvement, which is designed to upgrade bibliographic access to, and the physical quality of selected subject collections in, the stacks has been initiated recently with special grant moneys. The Project includes major funding for cleaning the stack collections. As a precursor to the larger project, cleaning of the Jewish Division collections on a volume-by-volume basis was undertaken in 1986, and was followed by the cleaning of STRC's collection of learned society publications, which is estimated to contain some 36,000 volumes. Cleaning is performed by an outside contractor using a five-member team; costs are 15 or 20 cents per volume, depending on the particular characteristics of the collections undergoing treatment. The Library's Development Office has been successful in attracting increased private funding for the collections maintenance and restoration programs managed by the Conservation Division. Conservation funding is a major component of the five-year Campaign for the Library, which seeks to raise $300 million; $4.1 million of the new money is being used to support various projects relating to preservation of the collections.

Another promising development in the area of increased funding is the recent establishment of a support group for the Conservation Division composed of Friends of the Library. The Lydenberg Society was established in 1986 and provides its members three programs a year on various preservation topics. Annual membership is $250 a year, and the proceeds are used to supplement institutional funds for the purchase of supplies, equipment and consulting services.

## *PRESERVATION THROUGH THE YEARS*

Beginning in the late 1930s and continuing to the present, microfilming has played a significant role in preserving science and technology materials for future generations. Since the filming program began, thousands of pamphlets, monographs and serials titles have been preserved by reformatting onto 35mm film stock and microfiche. The Second World War and its attendant defense and war production needs were an important stimulus to the microfilming effort.

In general, one can say that microfilming by libraries provides several advantages, among which are the following: (1) preservation of the intellectual content of publications, especially those printed since 1820, which are highly endangered and likely to disappear by the early decades of the twenty-first century; (2) significant space savings, especially for little-used materials; (3) a format that is inexpensive to produce copies from (once a title is preserved on a printing master, service copies can be produced "on demand" for as little as $12 per reel) and convenient to ship for purposes of interlibrary lending; (4) production of revenue from the sale of copies that can be used to fund additional preservation, or other library programs; (5) file integrity that is easily maintained; and (6) assuming that the master negative is produced and housed according to nationally-recognized standards, works preserved on film that will never go out of print. Weighing against these advantages are the high cost of establishing and staffing an in-house microfilming facility, the cost of purchasing microfilm reading devices and film storage units, and continuing public resistance to microfilm as an information storage medium. Many people continue to want to see and touch the original book.

In the past two years NYPL has filmed 431 science and technology titles amounting to more than 1,800 physical volumes. Among the titles filmed are *Drapers' Record* (London), 1889-1949; *Russell's National Motor Bus Guide* (Iowa), 1929-80; *Blast Furnace and Steel Plant* (Cleveland), 1916-71; *American Artisan, Tinner, and Home Furnisher* (Chicago), 1886-1971; *Die Chemische Industrie* (Berlin), 1878-1933; *Transactions of the North-East Institution of Engineers and Shipbuilders* (Newcastle-upon-Tyne), 1884-1951; and *Heat Treating and Forging* (Pittsburgh), 1917-56. Note: Titles filmed by NYPL are listed in the *National Register of Microform Masters; Microforms in Print*; and in *The New York Public Library Register of Microform Masters: Monographs* available on microfiche. Titles filmed since 1983, and selected titles filmed before 1983, are in the Research Libraries Information Network [RLIN] database. For further information on titles available and price information contact the Reprographic Services, New York Public Library, Fifth Avenue at 42nd Street, New York, NY 10018.

In addition to its own internal preservation microfilming program, NYPL has collaborated with facsimile reprint and micropublishers for many years. For example, The Research Libraries were a heavy contributor to Readex Microprint's *Landmark of Science Series*, Part I (Monographs) and Part II (learned Society Publications). Many additional titles were filmed in the late 1960s under a contractual agreement with the 3M Company's IM Press (no longer in business). The master negatives for those titles were retained by The Research Libraries, and service prints may be ordered through Reprographic Services.

It is important that libraries consider carefully the advisability of surrendering ownership of master negatives when approached by micro-publishers and other reprint firms wishing to gain access to materials in their collections. The vaults of micropublishers are filled with master negatives based on materials that were originally acquired, organized for use, and stored for many years by libraries at enormous expense. The cultural patrimony that libraries traditionally have held in trust, and which they hold responsibility for preserving, has in some instances passed into the hands of the for-profit sector, often for little more in return than a single microfilm copy—a pittance when one considers the long-term marketability of the product held by micropublishers.

Newly acquired science and technology materials received unbound are given full library binding treatment. Unfortunately, most materials must be oversewn because the printing and book manufacturing industry, for economic reasons, has largely abandoned issuing publications with intact folded signatures. Virtually all publications received today are already adhesive bound, leaving libraries with few options from which to choose when binding materials for long-term use and storage. Most of the patent materials received (the Patents Collection grows at a rate of 80 feet a year) are given storage binding rather than standard library binding. The cost difference between the two styles of binding is significant, one-third less. Storage binding, sometimes referred to as LUMSPECS binding after the committee of the American Library Association that developed *Minimum Specifications for Lesser Used Materials* for Libraries in 1959, consists of binders board cut flush to the text block with no cloth covering and no flyleaves, a cloth spine for labeling purposes, and reinforced inner hinges inside the front and back covers.

Beyond routine rebinding, little preservation work has been done on the monographs in the science and technology collections. Over the years, several hundred rare volumes with imprints prior to 1601 have been culled from the stack collections and stored in an area having limited access and provision for supervised reading. Having recently been thoroughly cleaned, these materials are now being examined on a title-by-title basis; most of them will be transferred to the Rare Book Room in the near future.

Conservation treatment, beyond routine rebinding, is costly, and until recent times the Libraries' resources have been insufficient to meet the need for careful and sensitive treatment of materials in the general collections that over the years have become scarce or even rare. Since 1984, however, each of the public service divisions, including Science and Technology, has been able to send carefully-selected materials to the Restoration Office for treatment on a regular monthly quota. Although the quota for science materials is small—fifteen hours of treatment time per month as of summer, 1986—the allocation is expected to rise in the future, as new staff is added to the Science and Technology Research Center. They will be trained to make appropriate preservation selection and treatment decisions. A similar quota system is in effect for materials requiring quick repair and/or routine rebinding by a commercial library binding firm.

The kind of treatment for book materials provided by the Restoration Office frequently entails taking a volume completely apart, washing, chemical stabilization (deacidifying) and air-drying of individual leaves, and reconstituting the leaves into signatures for resewing and binding into the original boards or into a sturdy library binding. Leaves that are extremely weak or brittle may be enclosed in polyester sleeves and the sleeves inserted into a post binding to maintain some semblance of the original codex format. Given the vast size and overwhelming needs of the stack collections and the small staff in the Restoration Office, the majority of treatments performed fall under the heading of "phased preservation," i.e., only basic treatment is undertaken at present aimed at stabilizing the object and protecting it from further damage, anticipating further treatment in the future as priorities and resources allow. For this reason, production of custom-fitted boxes is among the more important services provided at the present time by the Restoration Office for materials that have been identified as scarce or rare and which should be maintained in their original format.

In performing restoration work, careful attention must be paid to collation. In addition, the principle of reversibility must be observed. This doctrine holds that every form of treatment rendered by the conservator must be capable of being undone—i.e., reversed—with a minimum of effort by a conservator of the future, in case it is discovered that the first or the earlier treatment entailed use of materials that were later found to be unstable or harmful, or book structures were found in the long run to perform unsatisfactorily. A case in point is the rebinding of *Domesday Book,* in England's Public Records Office, done in the 1950s, which proved to be unsatisfactory and had to be rebound in 1986.

Food technology, cookery and cuisine is one of the collecting strengths of the Science and Technology Research Center, numbering in all about 80,000 titles. For the past three years STRC has been fortunate in having a librarian on the staff who is a specialist in the literature of food technology. Under his direction, a major menu collection amounting to some 25,000 pieces from famous restaurants in New York, other cities of the U.S. and Europe, assembled in the 1840s to the early 20th century, is being cleaned, inventoried and refoldered in acid-free containers by volunteers.

The single most effective measure that a library can take to further the long-term preservation of its resources is to assure adequate environmental conditions in all areas where materials are stored. NYPL has made progress in this area of collections maintenance, as manifested by the installation of an environmental control system serving the central stacks that became operational in 1985, but much remains to be done. A program is needed that will include systematic cleaning of the collections, identifying individual volumes in need of specific preservation treatment, maintaining (and enforcing) adequate standards of cleanliness in book storage areas, providing acceptable shelving units, training staff in safe handling practices, and providing for security from damage by fire and unforeseen water disasters such as follow upon faulty or improperly maintained plumbing systems. In this context, it is heartening to know that NYPL's security force has been augmented in recent years by the addition of personnel trained in fire prevention and disaster protection and recovery. A major stack refurbishment effort is in the planning stages that will extend the sprinkler system to all storage areas, and a disaster recovery plan has been designed and implemented. The plan includes the provisioning and stationing of mobile emergency disaster supply units in key locations throughout storage areas, and a communication system for rapid reporting of a disaster situation in order to assemble specially-trained personnel should the need arise. Despite the efforts, the Los Angeles Public Library disasters of 1986 cast a long shadow of apprehension. Our staff know all too well that "it could happen here."

## *PLANNING FOR THE FUTURE*

Over the past fifteen years a combination of adverse factors—primarily budget and space constraints—has resulted in deficiencies that threaten to undermine the public service and preservation efforts of the Science and Technology Research Center. Collection fragmentation, resulting in the removal of 60 to 70 percent of the collection to the Annex building, has created difficulties in selection of materials for preservation, and the situation has been exacerbated by staff shortages, which have resulted in little time being available for the important work of selection of materials for treatment.

In an effort to reverse this trend, The Research Libraries for the past three years have been assessing the Center's collections, services and preservation needs. It has been concluded that several programmatic measures are necessary to assure STRC's continuing vitality and future growth, and to enhance its usefulness to researchers and its importance as a key point of New York's network of educational resources for college, post-graduate and unaffiliated students, and business users. Increased preservation of materials stands high on the list of priorities for action. In order to undertake a comprehensive preservation program, an accurate assessment of the state of deterioration of the collections must be determined. This will require hiring a librarian specialist in science and technology with experience in preservation to work with staff of the Conservation Division for a period of one year to survey the collections. In addition, physical treatment and microfilming services must be increased over the next several years.

A three-phase plan has recently been approved in principle, to be undertaken over the next five years. The plan, which contains both short-range and long-range goals, addresses personnel needs, space requirements, and collection development and preservation needs. The long-range plan calls for removing those collections that remain in the Central Building to the Annex, and establishing at that site a new and revitalized Science and Technology Research Center, with the collections once again integrated. A Computer Science Resource Center and a Patents and Technical Information Unit are among the possibilities being discussed within the context of expanded services. Concurrently, preservation activities will continue and will be augmented. Significant grants announced in the past several months hold great promise for the future vitality and strength of the science and technology collections, with greatly improved housing, assessment of preservation needs, and provision of increased treatment services that will assure continuing public access to all of the materials in the collections.

# *NEW REFERENCE WORKS IN SCIENCE AND TECHNOLOGY*

Robert G. Krupp, Editor

*Reviewers for this volume are: Carmela Carbone (CC), Engineering Societies Library, New York, NY; Amy D. Cooper (ADC), University of Vermont, Burlington, VT; Kerry L. Kresse (KLK, University of Kentucky, KY; Robert G. Krupp (RGK), Maplewood, NJ; Barbara A. List (BAS), University of Michigan, Ann Arbor, MI.*

## EARTH SCIENCES

*Dictionary of rocks.* By Richard Scott Mitchell. New York: Van Nostrand Reinhold, 1985. 228 pp. $29.95. ISBN 0-442-26328-7.

The preface states that this is the first English dictionary dedicated to the names of rocks, and that the focus brings together those terms to form a comprehensive vocabulary. As such, this dictionary will have a definite utility for collections in this and related subject areas. The definitions are concise and indicate derivation of name, credit the individual who first used or defined it, and give the date of its introduction. The intended audience is broad, including laymen, amateur rock collectors, and students and professionals in petrology. Because some definitions rely on petrologic terms, a glossary has been included to aid the less sophisticated user. The following areas are within the scope of this book: igneous, sedimentary, and metamorphic rocks; migmatites, tektites, impactites, and major meteorite types; natural organic resins, ambers, bitumens, and major coal varieties; cave rocks and formations; and gem-rocks. Also included are non-English terms frequently found in English-language publications. The book does not cover minerals, soil science, petrology, coal constituents, and native crystalline organic compounds. Color plates and black and white photographs supplement the text. (BAL)

*Directory of geoscience libraries, United States and Canada*. 3d ed. Compiled by Nancy L. Crossfield (as editor) and others. Alexandria, VA: Geoscience Information Society (American Geological Institute), 1986. 99 pp. $20. ISBN not given.

A collection of information on 407 libraries covering geoscience literature, 80 more than reported in the 1974 2nd edition. Many are new and all entries possible have been revised. However, entries with an asterisk following the name of the parent organization represent collections deemed significant enough to be included, even though no direct response was received by the compilers. That includes 13 from the United States and 26 for Canada. Arrangement is alphabetical by state and then by Canadian province. There is also an organization index. This is a timely, comprehensive and relatively inexpensive directory, particularly useful for geoscience collections in industry, academe, government, and the larger public libraries.

*Challinor's dictionary of geology*. 6th ed. Edited by Antony Wyatt. New York: Oxford University Press, 1986. 374 pp. $15.95. ISBN 0-19-520506-5 (pbk).

Includes a selection of some 1,500 names and terms, a special feature being the copious quotations and references (to chiefly British geological literature). The definitions and explanations are generally 100-150 words but not a few are in the 300-500 word category (such as with "joint," "erratic," "silurian system," and "palaeomagnetism"). There is also a thorough classified index. For all geology collections despite the somewhat British bias. (RGK)

## ENGINEERING AND TECHNOLOGY

*Atlas of fatigue curves*. Edited by Howard E. Boyer. Metals Park, OH: American Society for Metals, 1986. 518 pp. $112.00. ISBN 0-87170-214-2.

A compilation of important and frequently referenced fatigue curves arranged by standard alloy designations and accompanied by a textual explanation and interpretation of test results. Almost 500 curves are provided. The original literature source is cited for each instance. For most engineering collections in industry and academe. (RGK)

*Corporate technology directory*. 1986 U.S. ed. 3 vols. Edited by Charles T. Peers, Jr. and Charles E. Caldwell. Wellesley Hills, MA: CorpTech, 1986. Mixed pagination. $650.00 ISBN 0-936507-03-9 (set).

This directory fills the need for a source of information on high-tech industry. It contains data on over 12,500 private and public United States and foreign high technology manufacturers operating in the United States. Included are manufacturers of computer hardware, software, photonics, robotics, artificial intelligence, biotechnology, and advanced materials. The directory also includes most public and major private manufacturers in aerospace, specialty chemicals, energy, factory automation, medical (sic), subassemblies, components, and telecommunications. There are no minimum size restrictions for inclusion. Thus, even the smallest start-up company can be listed. The firms are indexed by company name, and product type, by parent company, by location, and by executive name. Volume 1 is the Index volume; Volumes 2 and 3 contain the company profiles. (CC).

*(The) dBASE III programming handbook*. By Cary N. Prague and James E. Hammitt. Blue Ridge Summit, PA: TAB; 1986. 229 pp. $24.95. ISBN 0-8306-0676-9.

Actually this book is an advanced continuation of TAB's *Programming with dBASE III* and provides examples of algorithms for business use. However, technical subjects are treated on a very basic level. The author claims that ". . . the casual computer user has no real need to understand the intricate melding of programs with a system." Hardware and software in connection with this user-friendly database management system are introduced. For computer science collections dealing with the programming of microcomputers. (RGK)

*Encyclopedia of food engineering*. 2d ed. By Carl W. Hall and others. Westport, CT: AVI Publishing; 1986. 882 pp. $135. ISBN 0-87055-51-4.

This heavily revised encyclopedia continues to provide technical data relating to the application of modern engineering to the food processing industry. Emphasis is on equipment and facilities used in food handling, manufacture, and transportation. However, food and food products are not totally ignored. Many of the subjects cover 8-10 pages each but most are of the 1-2 page variety. References to sources of information are on the scant side. (RGK)

*Engineering formulas*. 5th ed. By Kurt Gieck. New York: McGraw-Hill; 1986. 544 pp. $19.95. ISBN 0-07-023231-8.

This is a revised and expanded English edition of a collection of technical and mathematical formulas. Included is a new section on statistics and addition of Fourier and Laplace transformations to the "arithmetic" making them available for personal notations. Flexible covers would not only make the volume easier to handle but also weigh less in the pocket for which it is designed. (RGK)

*Fachwörterbuch energie- und automatisierungs-technik.* (*Dictionary of power engineering and automation.*) Teil 2: Englisch/Deutsch. Edited by Heinrich Bezner. Munich: Siemens-Aktiengesellschaft; 1986. 471 pp. $49.95. ISBN 3-8009-1438-7. (Distributed by Wiley.)

A collection of 48,000 entries covering the fields of power engineering, electrical installation, and automation. Of special mention are topics such as switchgears, power electronics, and programmable controllers. A worthwhile addition to any technical translator's library of tools for translation needs of English to German. (RGK)

*Field inspection handbook.* Edited by Dan S. Brock. New York: McGraw-Hill; 1986. Mixed pagination. 464 pp. $49.50. ISBN 0-07-007932-3.

A comprehensive how-to guide which will assist in converting the design of a structure into a completed facility by providing the inspector with the engineering, technological, and practical guidelines to assure that the contractor's performance conforms fully and in detail with the intent of the designer. This reference work is for the resident engineer, inspector, or technician who may be a civil engineer. Well-indexed. Good illustrative matter. (RGK)

*(The) Foseco foundryman's handbook.* 9th ed. Rev. and edited by T. A. Burns. New York: Pergamon; 1986. 435 pp. $40.00. ISBN 0-08-032549-1.

This compilation is for those making castings by most of the usual routes except possibly investment casting which uses a rather special mould material. Special attention is given recent developments of new foundry work techniques. New areas reported in this edition include metal filtration and an extension to the specifications and grades of cast iron now commercially available. A concerted effort has been made to standardize on metric or SI units. There is a full index. (RGK)

*(The) handbook of chlorination.* 2d ed. By George Clifford White. New York: Van Nostrand Reinhold; 1986. 1070 pp. $89.50. ISBN 0-442-29285-6.

It is intended that this book be a handbook for the designer, a manual for the operator, a textbook for the student, and a guide for the regulatory agencies. It purports to bring together all the pertinent information necessary to provide a practical approach to the entire subject of chlorination as it is applied to potable water, wastewater, and industrial water. The second edition brings all the material up to date, but in order to keep the size of the book manageable, the following chapters in the first edition have been eliminated: Chlorination of Swimming Pools, Chlorination of Cooling Water, and Other Applications of Chlorine. (CC)

*Handbook of concrete engineering*. 2d ed. Edited by Mark Fintell. New York: Van Nostrand; 1985. 892 pp. $89.50. ISBN 0-442-22623-3.

This new edition (from the 1974 one) is considerably revised and includes new design techniques via use of computers, the 1983 version of the ACI Code, progress in construction equipment, and attention to higher strength materials. There are new chapters on slab systems, parking structures and structural plain concrete. Good documentation but with scant reference updating beyond 1974 for many of the chapters. Well-indexed. For civil engineers, architects, and contractors. (RGK)

*Handbook of corrosion resistant piping*. 2d ed. By Phillip A. Schweitzer. Malabar, FL: Krieger Publishing; 1985. 417 pp. $44. ISBN 0-89874-457-1.

This is a revision of a 1969 reference work designed to assist the engineer, specifier, and piping designer in the analysis of corrosion resistance and the costs of material installation and maintenance in the course of selecting optimal material of construction for piping systems in today's chemical, processing, and allied industries. The book is well-organized and indexed. For all serious chemical engineering collections in industry. (RGK)

*Handbook of electronic manufacturing engineering*. 2d ed. By Bernard S. Matisoff. New York: Van Nostrand; 1986. 564 pp. $52.50 ISBN 0-442-26072-5.

Addition of new material has heavily enhanced this edition for practicing manufacturing engineers who are involved in the production of electronic products from home entertainment to computers for high performance military aircraft and space vehicles. Excellent illustrative matter. A good introductory reference work to the field. Author affiliation not indicated. (RGK)

*Handbook of hydraulic resistance*. 2d ed., rev. and augmented. Edited by I. E. Idelchik. New York: Hemisphere; 1986. 640 pp. $89.95. ISBN 0-89116-284-4.

This is a translation from the 1975 Russian second edition and contains some 40% new and revised data by comparison with the first. Most chapters contain data on a definite group of fittings or other parts of pipelines and fluid network elements having similar conditions of liquid or gas motion through them. There are hundreds of illustrations of flow passages as well as scores of graphs. The documentation is extensive but only about 16% of the citations cover the six years (1970-1975) prior to publication of the Russian edition (or indeed even this translation); and none are listed thereafter. For engineering collections with a strong interest in hydrodynamics, especially as viewed by the Russian science community. (RGK)

*Handbook of modern electronics and electrical engineering*. Edited by Charles Belove. New York: Wiley; 1986. 2401 pp. $85.00. ISBN 0-471-09754-3.

It is impossible for an engineer to be expert in all the areas of engineering encountered in the course of an assignment. This had led to a constantly increasing need for up-to-date state-of-the-art information in the form of textbooks and handbooks. And this is especially true in electronics, a field that is used extensively in all branches of engineering. This reference work is designed to provide information to engineers working in fields other than electronics, to practicing electrical and electronics engineers, to management personnel, and to anyone else requiring such material. The 69 chapters in this handbook provide an overview of each subject area in reasonable depth, along with basic theory and design information. An extensive bibliography is included in each chapter. (CC)

*Handbook of public water systems*. Edited by Robert B. Williams and Gordon L. Culp. New York: Van Nostrand Reinhold; 1986. 1113 pp. $89.50. ISBN 0-442-21597-5.

This book provides a comprehensive source of information on all aspects of public water systems, from the point of source development to the delivery of the final product. It brings together information on historic practice and information from current research, evaluation, and application of new techniques for design and operation of public water systems. it is hoped that the material presented will be of interest to engineers, chemists, bacteriologists, treatment plant operators, utility managers, and administrators. It should also prove valuable as a reference for teachers and students interested in methods for enhancement of water quality. There are three useful appendices of conversion factors, chemicals used in the treatment of water and wastewater, and miscellaneous tables. (CC)

*(A) handbook of software development and operating procedures for microcomputers*. By Paul Holiday. New York: Macmillan; 1985. 181 pp. $24.95. ISBN 0-02-949510-5.

This book describes software development and operating procedures that can be used on personal computers and microcomputers using floppy disks or cassettes as mass storage devices. The subject level is intended to meet the needs of beginners and microcomputer novices. The author's main theme is *organization*, a vital concept in achieving systematic software and document development. None of the material should be beyond anyone who owns or has access to a microcomputer. Useful and extensive appendices are a glossary, procedure forms, and five program listings (e.g., BASIC) with examples. Author affiliation not given. (RGK)

*Heat exchanger sourcebook*. Edited by J. W. Palen. New York: Hemisphere; 1986. 805 pp. $59.95. ISBN 0-89116-451-0.

This sourcebook is a useful collection of practical articles by internationally known experts and industrial consultants in the field of process heat exchanger design and taken from four earlier (1974-1983) published volumes on the subject. The compilation is divided into six sections: General design information; Shell-and-tube heat exchangers; Reboilers and condensers; Plate heat exchangers; Heat exchange enhancement techniques; and Fouling. While the book does not constitute a comprehensive treatment of process heat exchanger design, it does present some of the most useful information available on the subject. (CC)

*(The) illustrated dictionary of microcomputers*. 2d ed. By Michael Hoordeski. Blue Ridge Summit, PA: TAB; 1986. 352 pp. $24.95. ISBN 0-8306-0488-X.

This is a revision of the 1978 *Illustrated dictionary of microcomputer terminology*. Both hardware and software terms are included. Over 8,000 terms are listed, a few with helpful diagrams or figures. For all computer science collections. (RGK)

*International directory of building research, information and development organizations*. 5th ed. Edited by G. Sebestyen and C. E. Pollington. London: E. & F. Spon Ltd.; 1986. 300 pp. $89.95. ISBN 0-419-12990-1.

This directory is sponsored by the International Council for Building Research, Studies and Documentation. It provides detailed information on over 600 research institutes, government organizations, universities and private companies in 54 countries. Each entry follows a standard pattern: after the address and telephone number of the institution, there is a brief description of its history and financial support followed by the names of the senior staff; total number of staff; the institution's structure and services; its main research programs; and a list of its publications. For this edition a subject index has been added allowing the reader to identify centers of research activity on individual construction topics throughout the world. The global span of the entries makes the directory an effective means for bringing together researchers and practitioners from various countries to address problems and research opportunities of common interest. (CC)

*(The) local network handbook*. 2d ed. Edited by Colin B. Ungaro. New York: McGraw-Hill; 1986. 389 pp. $30.95. ISBN 0-07-065917-6 (pbk).

A collection of articles on local area networking (LAN) citing relatively recent developments in areas such as technology, network design, implementation, planning and management, software, and applications. However, some of the reports are about PBX development (which is not LAN at all) but obviously will co-exist well with LAN. Advantages and disadvantages to both systems are presented. Recent cost reductions are also examined. For any library collection serious about computer networks and data communications. (RGK)

*Mechanical engineers' handbook*. Edited by Myer Kutz. New York: Wiley; 1986. 2316 pp. $79.95. ISBN 0-471-08817-X.

The editor's original intent was to provide a one-volume update of the two-volume *Kent's mechanical engineers' handbook*, 12th ed., 1950. However, this was not feasible because much modern material such as computers and nuclear power was not included and some of the older material such as the efficiency of splices and knots is still valid. The new handbook consists of chapters contributed by experts and organized in six parts: (1) Digital computers, (2) Materials and mechanical design, (3) Manufacturing engineering, (4) Systems, controls, and instrumentation, (5) Management and research, and (6) Energy and power. The section on manufacturing engineering includes discussions of computer-aided manufacturing and quality control. Finance and the engineering function, cost estimating, safety engineering, and sources of mechanical engineering information are included in the section on management and research. Solar energy, geothermal resources, cogeneration, nuclear power, cryogenic systems, and the impact of chemical and thermal processes on the environment are treated in the energy and power section. (CC)

*1986 international satellite directory*. Corte Madera, CA: Design Publications; 1986. Mixed pagination. $225.00. ISBN 0-936361-01-8.

This is the first edition of what is to be an annual guide to the international satellite industry. Included are directories on international organizations where information on members range from extensive, as with INTELSAT, to meager, as with INTERSPUTNIK. In another chapter there is heavy coverage of ground satellite equipment and lesser coverage of space and interface equipment. There is also extensive coverage of users and providers of satellite services, listings of eleven kinds of services to the satellite industry (e.g., research centers and consultants). A most interesting and detailed section deals with identification and orbiting maps of geosynchronomous satellites. This directory will find good use in any technical collection on satellites and even in larger public libraries. (RGK)

*Piping stress handbook.* 2d ed. By Victor Helguero M. Houston, TX: Gulf Publishing; 1985. 375 pp. $67. ISBN 0-87201-703-6.

A reference work to provide formula, technical data and other pertinent design information not always readily available for the piping stress/analyst in the petrochemical industry. New material in this revised edition includes piping branch reinforcements and stiffness coefficients for nozzles on cylindrical vessels as well as the latest ANSI/ASME Piping Codes. The bulk of the contents is in tabular form. The author is with S.P.I. Engineering, Houston. (RGK)

*Quality technician's handbook.* By Gary Griffith. New York: Wiley; 1986. 532 pp. $45.95. ISBN 0-471-82258-2.

This is a handy, practical guide designed to aid operators, machinists, and inspectors who have had unidirectional experience or want to study the mechanical trades for the first time in connection with producing, inspecting, and controlling quality products. A large portion of the work is rather elementary as with mathematics, blueprints, measuring tools, basic probability, and statistical process control. Best for personal use in machine-shop practice where there is high attention to quality control. (RGK)

*Robotics technical directory 1986.* Edited by William M. Rowe. Research Triangle Park, NC: Instrument Society of America; 1986. 170 pp. $49.95. ISBN 0-87664-917-7.

The General Information section of this directory lists in table form the typical characteristics of robot types, applications and reasons for use, industrial robot characteristics, and costs for typical robot systems. The Consultants section, besides names, addresses and telephone numbers, includes brief descriptions of specific expertise and experience. The Industrial Robots-Technical Information section is alphabetical by manufacturer. This listing includes descriptions, technical information, and drawings or photographs (where available) for industrial robots. There are sections that list manufacturers, suppliers and distributors of robots and robotic systems, manufacturers of automated factory equipment, and products and services that related to the robotics industry. The next three sections list companies and universities involved in robotics research, periodicals aimed specifically at the industry, and associations for the study and advancement of the technology. The directory includes a glossary of robotic terms and an index. (CC)

*Systems troubleshooting handbook.* Edited by Luces M. Faulkenberry. New York: Wiley; 1986. 415 pp. $44.95. ISBN 0-471-86677-6.

This is an introduction for the novice troubleshooter and specifies the tools and techniques used to identify malfunctions in electronic systems. The work may also aid more experienced troubleshooters in solving problems outside their area of specialization. There are three sections: test equipment and troubleshooting basics, analog system troubleshooting, and digital system troubleshooting. Well-illustrated and indexed. And there are appendices on microprocessor and robotic basics. (RGK)

## HEALTH SCIENCES

*Biotechnology directory 1986: Products, companies, research and organizations.* 3d ed. By J. Coombs, New York: Stockton Press; 1986. 518 pp. $155 (pap). ISBN 0-943818-20-6.

Over 1000 changes have been incorporated in this new edition, though scope and layout remain as used previously. It is still an annual reference book which presents information on biotechnology from a wide variety of sources. The term biotechnology, as it is used here, covers topics from industrial microbiology, genetic engineering, waste treatment and tissue cultures. Among the entries are professional organizations, government agencies, private companies, research institutes, and academic departments. Except for the chapters on international organizations and information sources (journals and indexing/abstracting services), each section is arranged by country, with indexes referring the reader to specific entries. Academic and special libraries will find this unique volume most useful. (KLK)

*Cancer rates and risks.* 3d ed. Compiled by Harriet S. Page and Ardyce J. Asire. [Bethesda, MD?]: U.S. Dept of Health and Human Services, Public Health Service, National Institutes of Health; 1985. 136 pp. $5.00. ISBN not available. (NIH publication no. 85-691.) (Order from: C. W. Associates, P.O. Box 34099, Bethesda, MD.)

This brief work is an update of the 2d edition, 1974. It answers many of the statistical questions which arise so regularly in libraries. The work is in two sections. Part 1, "Rates," provides statistics on cancer death rates and incidence in twenty countries. The remainder of part 1 is devoted to United States cancer incidence and cancer death rates. Part 2, "Risks." summarizes current knowledge of such general risks as alcohol, diet, and tobacco. This is followed by a discussion of risks for the most common cancers. A brief glossary and a nine page bibliography are included in this very useful work. Recommended for all types of health science libraries. (ADC)

*(The) Charles Press handbook of current medical abbreviations.* 2d ed. Philadelphia, PA: Charles Press; 1984. 180 pp. $9.95. ISBN 0-914783-00-9.

A-Z listing of medical abbreviations currently used in clinical medicine. Parenthetical notes define abbreviations with multiple meanings and those derived from Latin. Four pages of symbols complete this useful and reasonable book. Recommended for academic health science and hospital libraries of all sizes. (ADC)

*(The) dictionary of vitamins: the complete guide to vitamins and vitamin therapy.* By Leonard Mervyn. New York: Thorsons; 1984. 190 pp. $12.95. ISBN 0-7225-0869-7 (pbk.).

This dictionary includes entries for vitamins, vitamin deficiency diseases, foods containing vitamins, and leading researchers in the field of vitamins. The vitamin definitions provide thorough coverage of food sources, importance for health, diseases caused by deficiency, and recommended daily intakes. Other entries are brief. The target audience for this clearly written dictionary is the layperson. Recommended for public and academic libraries. (ADC)

*Encyclopedia of medical history.* By Roderick E. McGrew. New York: McGraw-Hill; 1985. 400 pp. $34.95. ISBN 0-07-045087-0.

Provides access to the historical aspect of major medical subjects. Each entry gives chronological coverage of the topic. There are no biographical articles, but persons are discussed as part of a topic and these personal names are included in the index. A bibliography of supplemental reading follows each article. Intended for students of both history and medicine. Recommended for larger public libraries, general academic libraries, and medical libraries. (ADC)

*(The) essential guide to generic drugs.* By M. Laurence Lieberman. New York: Harper & Row; 1986. 326 pp. $8.95. ISBN 0-06-096040-X (pbk.).

This guide provides an alphabetical listing of drugs by generic name. There are cross-references by brand name. For each drug, the guide answers the following questions: Does the generic drug offer a savings? Is the brand name drug a better choice? Is the generic drug FDA-approved as bioequivalent to the brand name? Would a switch from brand name to generic (or from one brand to another) be harmful? Also, includes information on drug regulations and state generic drug laws. Recommended for public libraries, health science libraries, pharmaceutical libraries, and personal purchase. (ADC)

*(The) guide to the nation's hospices.* 1986 ed. Arlington, VA: National Hospice Organization; 1986. 139 pp. $47.00. ISBN 0-931207-01-0.

The guide was produced from a survey conducted by the National Hospice Organization, with some information from a study by Children's Hospice International. The two surveys identified 1,465 hospices currently providing services and 103 in planning phases. The directory lists 1,418 programs. Hospices are arranged alphabetically by state and city. Entries are brief. For each hospice, the following information is included: name, address, contact person, phone number, type of program, and scope. It is intended to be updated annually. Recommended for larger public libraries and health science collections. (ADC)

*Handbook of injectable drugs.* By Lawrence A. Trissel. 4th ed. Bethesda, MD: American Society of Hospital Pharmacists; 1986. 650 pp. $46.22. ISBN 0-930530-62-4 (softcover).

The *Handbook* provides data on injectable drug stability and compatibility for 242 commercially available drugs and 60 investigational drugs. The fourth edition retains the same format as previous editions, with information updated, and the addition of 29 new commercial drugs and 14 new investigational drugs. Commercial drugs are in alphabetical order by nonproprietary name. Original sources are referenced. American Hospital Formulary Service classification numbers are included so that therapeutic information can be located. For each drug, the following is provided: concentration, stability, pH, dosage, and compatibility information. A useful reference source for biomedical, large hospital and pharmaceutical libraries. (ADC)

*International dictionary of medicine and biology.* Edited by Sidney I. Landau. New York: John Wiley & sons; 1986. 3 vol. $395.00. ISBN 0-471-01849-X.

This scholarly dictionary was 10 years in the making and contains more than 159,000 concise definitions of over 151,000 terms. It is very broad in scope, covering the traditional areas of basic and clinical medical science, plus specialties involving new technologies and the delivery of health care systems. Thus the reader will find the language of molecular medicine along with terms of historic significance. Preference is given to American spellings, with cross-references from British spellings when necessary. Definitions include etymologies, variant spellings, and usage information. Over 80 advisory editors and 90 contributors generated the vocabulary and definitions within their specialties. Main entries are arranged alphabetically letter-by-letter. As the most comprehensive and current dictionary of its kind, it will be a valuable addition to most medical and life science collections in spite of its high cost. (BAL)

*Mosby's medical & nursing dictionary.* 2nd ed. Edited by Walter D. Glanze and others. St. Louis: Mosby; 1986. 1563 pp. $21.95. ISBN 0-8016-5195-6.

This second edition has 3,000 new entries and incorporates two new categories: dentistry and computer technology. The fields of geriatrics and radiology are given better coverage. Metric equivalents are provided.

There's a new pronunciation system (that used by many college dictionaries). Definitions are clearly written. At the front of the dictionary is a 44-page color atlas of human anatomy. The appendices contain sixteen sections with such topics as medical terminology, DRG's, laboratory values, and DSM-III classification. This is a good, current dictionary. Recommended for health science libraries. (ADC)

*(The) Oxford companion to medicine*. 2 vols. Edited by John Walton and others. New York: Oxford University Press, 1986. 1524 pp. $95.00. ISBN 0-19-261191-7 (set).

This fine Oxford reference work is almost an encyclopedia of medicine. Entries range from short definitions of terms to articles of up to 10,000 words. The lengthier articles are on major disciplines, medical specialties, and other important topics, e.g., biochemistry, cardiology, surgery. Biographical entries for medical notables are included. (No one still living is included.) The work is intended to be used by medical students and practitioners' nurses, and other health professionals. It can also be understood by the layperson. There are "see" and "see also" references and words used in one entry which are entries themselves are indicated. Recommended for academic health science libraries. (ADC)

*Physics in medicine & biology encyclopedia: medical physics, bioengineering and biophysics*. Edited by T. F. McAinsh. New York: Pergamon Press, 1986. 2 volumes. $185.00. ISBN 0-08-026497-2.

This encyclopedia is meant to serve as a handbook and guide to the impact physics had made in the areas of biology and medicine. It appears to give greater emphasis to the needs of medical professionals than it does the pure biologists, and its intended audience is described as hospital physicists, medical technicians, and clinicians. There are over 200 articles written by experts. The reader is expected to have some basic understanding of physics and at the same time, may have little familiarity with the life sciences. Topics that would call for a "highly mathematical treatment" have been avoided. A nice feature of these volumes is the classified list of articles, which groups all of them into 25 broad topics. If an article falls in more than one topic, it is listed under the additional areas. This makes it convenient for the reader to consult coverage of an area of interest, such as the neurological sciences. A small sampling of subjects treated include nuclear medicine, lasers, genetic engineering, membrane physics, radiotherapy, and radiobiology. The signed articles, which vary in length from 1 to 7 pages, include bibliographies and references that will lead the reader to further sources. A glossary is provided to aid with terms associated with anatomy, physiology, and pathology. (BAL)

*Remington's pharmaceutical sciences*. 17th ed. Edited by Alfonso R. Gennaro. Easton, PA: Mack Pub. Co., 1985. 1984 pp. $85.00.

This book is a classic. The first edition was published in 1886 and, beginning with the thirteenth edition, in 1965, it has been published every five years. The work started as a practical manual for the retail pharmacist and has evolved into a comprehensive treatise on the theory and practice of the pharmaceutical sciences. The organization remains the same as in recent previous editions with division into nine parts such as "Pharmaceutical Chemistry," "Testing and Analysis," "Pharmaceutical and Medicinal Agents" (the largest section which includes an extensive list of drugs, each of which is thoroughly described). The drug monographs are grouped by function and have been thoroughly updated. There are chapters on drug interactions, drug literature, and laws governing pharmacy. This is an essential text for biomedical and pharmaceutical libraries. (ADC)

## LIFE SCIENCES

*American insects: a handbook of the insects of America north of Mexico*. by Ross H. Arnett, Jr. New York: Van Nostrand Reinhold; 1985. 850 pp. $79.50. ISBN 0-442-20866-9.

While this volume will certainly be used by professionals, it has been designed to serve the nonspecialist engaged in the quest to identify and access information on insects of the United States and Canada. It gives information on sizes, shapes, color patterns, and important features of representative species for each major family likely to be encountered by the general collector. Bypassing obscure groups, it lists each order, keys each family, and lists each genus and the number of known species in the geographic area covered. In addition, the author provides general bibliographic references to all taxon, illustrations for most of the features used in the keys, and general biological information and distribution for the taxa covered. Highly technical terms are avoided when possible, although a glossary is there to assist when needed. While most of the book is devoted to a detailed description of the insect orders and species, introductory chapters offer more general discussions on classification and systematics, identification features, ecology, behavior and distribution, and collecting insects. (BAL)

*Atlas of the rabbit brain and spinal cord*. By J. W. Shek and others. Basel; New York: Karger; 1986. 139 pp. $198. ISBN 3-8055-3814-6.

The authors, researchers at the New York State Office of Mental Retardation and Developmental Disabilities, point out that this is the first atlas that includes coverage of both the rabbit brain *and* spinal cord. There are more than 300 illustrations, only a few of which are line drawings. The rest are high quality black and white photographs, reproduced on a heavy, glossy stock. All sections of the brain are illustrated here, as well as transverse sections at 32 levels of the spinal cord. The section on materials and methods provides a detailed account of how the rabbits were prepared, dissected and then displayed. Note that this atlas is not intended for stereotaxic purposes. The book is bound well to last through heavy use. Researchers in pharmaceuticals, veterinary medicine and related biomedical disciplines, both in academia and industry, will find this reference manual most helpful. (KLK)

*Birds of New Guinea*. By Bruce M. Beehlr and others. Illustrated by Dale A. Zimmerman and James Coe. Princeton, NJ: Princeton University Press; 1986. 293 pp. $37.50. ISBN 0-691-08385-1 (alk. pap.).

The lush tropical environment of equatorial New Guinea supports a fantastic array of bird life. In this field guide 725 species of birds are documented. Species summaries include physical description, voice, habitats, range, and also mentions similar species. More than 600 of the described species are illustrated in 55 exceptional quality plates, most of which are in color. Skins and photographs loaned by the American Museum of Natural History provided much of the detail for the illustrations. In addition, the authors summarize geographic, climatological and ornithological features of New Guinea, the world's second largest island. English names are used throughout the text, due in part to the unexpected variations of native names and languages. A very strong index lists the English names and also taxonomic names. This unique work is highly recommended for academic collections, as well as natural history collections in special libraries. Also recommended as a personal purchase for specialists. (KLK)

*(The) Cambridge encyclopedia of life sciences*. Edited by Adrian Friday and David S. Ingram. Cambridge and New York: Cambridge University Press, 1985. 432 pp. $39.95. ISBN 0-521-25696-8.

This will be a useful book to have on hand in reference collections for the synthesis it provides of modern biology. Working within the realm of the naturalist tradition and the findings of experimental research, it conveniently organizes information that normally would appear in many different sources. Starting at the level of the cell, the discussion covers the principles of growth, development, physiology and reproduction, and moves on to behavior and ecology. The next section looks at environments, focusing on the physical and climatic features of marine, coastal, terrestrial, freshwater and wetland environments. It also considers the living organism itself as an environment. The third and last section surveys evolution and the fossil record. A particularly valuable feature of this volume is the list of further readings which appears at the end of each chapter. They tend to be references to standard texts that one would expect to find in collections covering the life sciences. The book is generously illustrated with maps, tables, photographs and drawings. (BAL)

*(The) Collinridge dictionary of plant names.* B. Allen J. Coombes. Middlesex, England: Collingridge, 1985. 207 pp. $8.98. ISBN 0-600-35770-8.

The stated aim of this book is to "provide a guide to the derivation, meaning and pronunciation of the scientific names of the more commonly grown plants." The information it contains is presented in a very concise manner. Organization is alphabetically by generic and common names. The name of each genus is followed by the suggested pronunciation, its family, and derivation of name. This is followed by a short comment describing main use of the plants in gardens, and an indication of hardiness. After the above is compared, the species within each genus are listed alphabetically. Again, there is a guide to suggested pronunciation which precedes the meaning of the name, the common name, and the country of origin. Common names are cross-referenced to scientific names. Readers will find a short glossary to help with any technical terms that occur. (BAL)

*Conifers.* Text by D. M. van Gelderen; photographs by J. R. P. van Hoey Smith. Published in cooperation with Royal Boskoop Horticultural Society. Portland, OR: Timber Press, 1986. 375 pp. $65.00. ISBN 0-88192-056-8.

This volume, which was published on the occasion of the 125th anniversary of the Royal Boskoop Horticultural Society, The Netherlands, contains color photographs of over 1000 conifers of the class Coniferopsida and the class Taxopsida. Because this book is aimed toward readers with an interest in ornamental horticulture, each genera is represented in proportion to its importance to that field. Each genera that has been included is described in terms of the species within it, physical descriptions, habitat, and fruit. An emphasis has been placed on describing and illustrating specific characteristics of the genera treated. The appendices at the back of the volume include information on hardiness zones, systematics of modern Gymnosperms, botanical terms, plus a key to the coniferous genera. The book excludes coverage of the genera *Ginkgo, Ephedra,* and *Welwitschia*. (BAL)

*Eastern forests.* By Ann Sutton and Myron Sutton. New York: Alfred A. Knopf, 1985. 638 pp. Paperback: $14.95. ISBN 0-394-73126-3. (The Audubon Society Nature Guides.)

This field guide covers eight forest types found in North America: boreal, transition, mixed deciduous, oak-hickory, southern Appalachians, pine barrens, southern pinelands, and subtropical. Discussions in each case emphasize habitats, relationships among the plants and animals found there, and special features a visitor would want to see. Included with each article is a list of the principal trees, birds, butterflies, moths, insects, spiders, wildflowers, reptiles, amphibians, mushrooms, and mammals found in that forest type. Each item in the lists is numbered, enabling the reader to move easily into the color plates or the species descriptions. The plates depict woodlands and forests, and over 600 plants and animals. Species descriptions provide data on range and habitat, along with any other important characteristics of the plant or animal. The three appendices are made up of a glossary, a bibliography, and an index. (BAL)

*(The) encyclopedia of birds*. Edited by Christopher M. Perrins and Alex L. A. Middleton. New York: Facts on File Publications, 1985. 447 pp. $35.00. ISBN 0-8160-1150-8.

This volume will be a valuable addition to collections catering to either the amateur or the expert ornithologist. Much of its value comes from the provision of information recently uncovered about birds, particularly in regards to the behavior of tropical species. Beyond currency, it offers exceptional photographs and color and line drawings that emphasize behavior and physical attributes. The book is organized taxonomically, with individual articles treating single families or groups of closely related families. Articles, contributed by experts, cover physical features, distribution, evolutionary history, classification, breeding, diet and feeding behavior, social dynamics and spatial organization, conservation and relationships to man. Maps illustrate their natural distributions and show both breeding and wintering grounds for migratory birds. When research has uncovered particularly noteworthy behavior in a species, the article devotes a page or two to discussing it. The subject index lists both common and scientific name. (BAL)

*(A) field guide to western reptiles and amphibians: field marks of all species in western North America, including Baja California*. Text and illustrations by Robert C. Stebbins. Sponsored by the National Audubon Society and National Wildlife Federation. Boston: Houghton Mifflin Co.; 1985. 336 pp. $17.95. ISBN 0-395-08211-0. (Paperback: $12.95. ISBN 0-395-19421-0.) (The Peterson Field Guide Series; 16.)

This revised edition of Stebbins' 1966 field guide to western reptiles and amphibians supplements *Reptiles and Amphibians of Eastern and Central North America* by Roger Conant. As such, it covers western North America including Baja California as far east as the eastern boundaries of New Mexico, Colorado, Wyoming, Montana, and Saskatchewan north to the Arctic Circle. A total of 244 species along with 260 subspecies are described. A very important feature to this field guide is the high quality of illustration. Stebbins has provided 48 plates, thirty-five of which are in full color. Of the 244 species covered, 239 are illustrated. In addition, there are 200 distribution maps. All in all, there are 601 illustrations to aid the reader in identifying animals in the field, the hallmark of a first-class field guide. New information in this edition includes reproduction data for reptiles: clutch size, frequency of nesting, and time of laying. The species accounts give detailed physical descriptions, plus information on behavior, habitat, and similar species. The identification sections cover key features of color and structure in adult animals. A glossary is provided for the few technical terms that are used. (BAL)

*(A) field manual of the ferns and fern-allies of the United States and Canada*. By David B. Lellinger. Washington, DC: Smithsonian Institution Press, 1985. 389 pp. $45.00. ISBN 0-87474-602-7. (Paperback: $29.95. ISBN 0-87474-603-5.)

Mr. Lellinger, Curator of Ferns at the Smithsonian Institution, has provided keys to enable identification of 406 species, subspecies, and varieties of ferns and fern-allies that are native to or naturalized in Canada and the United States, excluding Hawaii. Most of the represented species are illustrated by a color photograph that either depicts them in their habitat or reveals particularly important identifying characteristics. The text is quite technical and is well-served by the glossary of terms which even includes drawings to enhance clarity. The key is followed by a bibliography and indexes to both common and scientific names. The novice will find the introductory chapter helpful, since the author covers the study and collection of ferns, an explanation of names and classification, notes on geography, habitats and ecology, a description of morphology and anatomy, and a summary of life cycle. (BAL)

*Glossary for horticultural crops*. By James Soule. Sponsored by the American Society for Horticultural Science. New York: Wiley, 1985. 898 pp. $42.50. ISBN 0-471-88499-5.

In compiling this glossary, the author aimed to cover both the traditional terms of horticulture and the newer ones in plant-related sciences such as genetic engineering and molecular biology. He has, therefore, taken a fairly broad view of horticulture. His intended audience includes students, teachers, researchers, extension specialists, administrators, growers, and anyone else involved in the culture of plants. The terms are arranged alphabetically within six sections, and are then cross-referenced and indexed by term and by crop. The sections cover horticultural crops; morphology and anatomy; horticultural taxonomy and plant breeding; horticultural physiology and crop ecology; propagation, nursery handling, soils, and crop production; and postharvest handling and marketing. Line drawings appear throughout and do a nice job of supplementing the text. Nearly 14 pages of selected references are provided at the back of the volume. (BAL)

*Guide to the birds of Colombia*. By Steven L. Hilty and William L. Brown. Illustrated by Guy Tudor. Princeton, NJ: Princeton University Press; 1986. 836 pp. $42.50. ISBN 0-691-08371-1 (alk. pap,).

The jungles of South America are rich in animal life, especially the areas surrounding the Amazon. The varied geography and climates also add to the incredible diversity. In testament to this, nearly 1700 bird species are chronicled in this excellent reference work. The information provided is straightforward and complete, including physical description, voice, breeding, habitat, range, and behavior. An interesting and important addition is the listing of similar species. The sixty-nine color and black and white plates, most of which are by Guy Tudor, are excellent and illustrate hundreds of the described species. Additional black and white line drawings are scattered throughout the text. The appendices offer advice on finding birds, an extensive bibliography, range maps, and finally, indexes to English names, genera and species. Introductory chapters summarize the unique topography, climate, vegetation and ornithology of Colombia. A fine purchase for academic libraries, museum or natural history collections, as well as personal collections. (KLK)

*(The) literature of the life sciences: reading, writing, research*. By David A. Kronick assisted by Wendell D. Winters. Philadelphia: ISI Press; 1985. 219 pp. $29.95. ISBN 0-89495-045-2.

As the author states in his preface, this book is meant to "provide a general background of observations and useful information for the practitioner, investigator, and student in the life sciences, so that they will be able to read, write, and research the literature in a more perceptive and efficient manner." Written from an historical viewpoint, it is neither a comprehensive guide or an inventory of sources, although the reader will find a selected list of guides to the literature in the appendix. The 12 chapters cover issues pertinent to the information systems in chemistry, biology, and medicine: developmental history, primary and secondary sources, characteristics of the literature, evaluation of the literature, indexing languages, citation indexing, personal information files, etc. The list of 484 references will also be of interest. Librarians, in addition to the primary intended audience, will find that this work offers a helpful perspective when faced with the monumental task of coping with the life sciences literature. (BAL)

*Marine fauna and flora of Bermuda: a systematic guide to the identification of marine organisms*. Edited by Wolfgang Sterrer in cooperation with Christiane Schoepfer-Sterrer. New York: Wiley, 1986. 742 pp. $99.95. ISBN 0-471-82336-8.

Designed for use by both laymen and scientists, this identification guide treats over 1500 selected species found in Bermuda's ocean waters. It covers plants and animals that, among other criteria, are large, conspicuous and abundant, and live near to the shore and close to the water's surface. Other species that are the only representatives of major taxa are also included. The point is made that much of Bermuda's marine life overlaps with that of the West Indies and the southeastern United States, thereby widening the applicability of this volume. The text is arranged hierarchically to reflect evolutionary development and the relatedness of organisms, and each section has been written by a specialist in the field. The 2870 illustrations (both black and white drawings and color plates) meticulously represent the species under discussion, and complement the discussions very nicely. The reader will also find over 800 references, a glossary, and a taxonomic index, all adding to the utility of this work. (BAL)

*(The) world of butterflies*. By Valerio Sbordoni and Saverio Forestiero. Translated from the Italian 1984 edition by Neil Stratton, Hugh Young, and Bruce Penman. Dorset, England: Blandford Press; 1985. 312 pp. $28.00. ISBN 0-7137-1500-6.

Although this book has been compiled to inform general readers about the world of the Lepidoptera, it should prove to be a book that will receive steady use if included in a natural as well as general science reference collection. The text is organized into 16 main chapters that cover such topics as structure, origin, and relationships of butterflies and moths; life cycle and metamorphosis; origin of species; classification and phylogenetics; behavior; geographical distribution; and more. The text is beautifully supplemented with color drawings (as opposed to the current trend of using elegant photographs) that serve to illustrate the discussion at hand and link with the great natural history illustrations of the last century. The authors have placed an emphasis on the diversity shown by butterflies and moths, and how that diversity can be explained in terms of adaptation and evolution. Special behaviors such as mimicry and migration are also covered. In terms of nomenclature, the authors have used the most widely adopted names. They do not give author name with the taxon. A detailed subject index is included. (BAL).

## PHYSICAL SCIENCES

*Atlas of historical eclipse maps. East Asia 1500 BC-AD 1900*. By F. R. Stephenson and M. A. Houlden. New York: Cambridge University Press, 1986. 431 pp. $79.50. ISBN 0-521-26723-4.

This is a map collection of solar eclipses (only) as bequeathed to mankind by Chinese astronomers. On each of the eclipse maps there is marked the site of a major Chinese city, usually the capital of the time, as a reference point. Main emphasis is on accurate mapping of the paths of actual eclipses. However, this atlas is by no means complete for partial eclipses visible in the Far East. A total of 862 maps are provided, two to a page. In addition to astronomy observatory library collections, volumes belong in academe and the larger public libraries. (RGK)

*Atmospheric chemical compounds: sources, occurrence, and bioassay*. By T. E. Graedel and others. New York: Academic Press; 1986. 732 pp. $55. ISBN 0-12-294485-2.

Primarily this reference work is about the chemical compounds found in the earth's atmosphere. Groupings are by chemical structures and properties of the compounds. There is also provided a perspective on these extensive data, initially with overviews of the properties of the earth's atmosphere, the indoor atmospheric environment, and generic toxicology. Thousands of chemicals and their properties are cited from over 1400 literature references. For industrial libraries with a high interest in the techniques of environmental chemistry. The first author is with AT&T Bell Laboratories. (RGK)

*(The) chemistry of organic selenium and tellurium compounds*. Vol. 1. Edited by Saul Patai and Zvi Rappoport. New York: Wiley; 1986. 939 pp. $271. ISBN 0-471-90425-2.

This is the first in a two-volume set covering all important aspects of the organic chemistry and the derivatives of selenium and tellurium. The contributors have, where possible, made comparisons between analogous compounds containing the three chalcogen atoms (from Group VI in the Periodic Table), sulfur, selenium, and the tellurium. By and large literature coverage is up to the end of 1983, with occasional references from 1984 (altogether some 4,3000 citations). Note that the material in this set falls outside the scope of the set of four volumes in the same series, entitled *The chemistry of the metal-carbon bond*. Well-indexed by author and subject. For all comprehensive chemistry collections. Editors with The Hebrew University, Jerusalem. (RGK)

*Concise dictionary of chemistry*. Edited by John Daintith. New York: Oxford University Press; 1985. 308 pp. $16.95. ISBN 0-19-866143-6.

This dictionary is derived from the *Concise science dictionary* of 1984. It consists of all the entries relating to chemistry, including physical chemistry and many terms used in biochemistry. Some 3,000 terms are listed. For physical science collections in undergraduate universities and larger public libraries. (RGK)

*Concise dictionary of physics*. Edited by Alan Isaacs. New York: Oxford University Press; 1985. 295 pp. $16.95. ISBN 0-19-866142-8.

This dictionary is derived from the *Concise science dictionary* of 1984. It contains all entries relating to pure physics plus some relating to astronomy, physical chemistry, mathematics, metal science, computing, and electronics as involved with physics. About 3,000 terms are provided. For physical science collections in undergraduate universities and larger public libraries. (RGK)

*Condensed imidazoles: 5-5 ring systems*. By P. N. Preston. New York: Wiley; 1986. 411 pp. $125. ISBN 0-471-88384-0. (The chemistry of heterocyclic compounds v. 46.)

> This volume contains a survey of 51 ring systems in which an imidazole ring is fused to an additional five-membered ring system. The synthesis, physicochemical properties, and reactions of compounds in each ring system are covered; reactions are organized on a mechanistic basis and the survey includes compounds that are both partially and fully saturated. This work should encourage study of the bicyclic condensed imidazoles and related compounds containing three or more rings. Literature coverage is through Volume 97 of *Chemical Abstracts* (756 citations involved). For extensive organic chemistry reference collections in industry and academe. (RGK)

*Data for biochemical research*. 3d ed. By Rex M. C. Dawson and others. Oxford, England: Clarendon Press; 1986. 580 pp. $59.00. ISBN 0-19-855358-7.

> Seven years have passed since the publication of the second edition. This new edition is intended to provide information on chemical compounds that a biochemist would be very likely to need in a laboratory setting. Toward that end, the authors state that the book is not comprehensive and does generally only include data that has the highest potential for use by the widest range of all biochemists. The main body of this work consists of 25 tables that arrange compounds in functional groupings as in, for example, amino acids, amines, amides, peptides, and their derivatives; vitamins and coenzymes; pharmacologically active compounds; and so on. Within each table compounds are arranged alphabetically. Examples of data given for each compound included synonyms, molecular formula, molecular weight, physical properties such as boiling-point or melting-point, and solubility. References to relevant literature on methodology are given when the authors judged them be helpful for the reader. Those who have used the previous editions will find some material deleted in favor of new data reflecting advances in biochemistry and molecular biology. While the book is directed toward the scientist's laboratory, it should prove to be a very handy reference source for chemistry and biology library collections. (BAL/KLK)

*Handbook of chemicals production processes*. Edited by Robert A. Meyers. New York: McGraw-Hill; 1986. Mixed pagination. $69.50. ISBN 0-07-041765-2.

> Forty important technologies for the production of major organic chemicals, inorganic chemicals, and polymers, prepared by the process licensors, are presented in this handbook. It should be noted that some of the requested information may not be disclosable and hence was omitted. This is the third in the Chemical Process Technology Handbook Series (the first is *Handbook of synfuels technology*; the second *Handbook of petroleum refining processes*). All three thus cover the production of the major feedstocks used for the synthesis of fuels, chemicals, and polymers and the production of these

feedstocks from alternative fossil fuels as well as the production of chemicals and polymers themselves. Contributions in this final volume are from nineteen firms located in the United Kingdom, the Federal Republic of Germany, Japan, The Netherlands, and the United States. For all major library collections on industrial chemistry. (RGK)

*Handbook of coordination catalysis in organic chemistry*. By Penny A. Chaloner. Boston: Butterworths; 1986. 1002 pp. $110.00. ISBN 0-408-10776-6.

This work demonstrates the increasing importance of homogeneous catalysis by metal complexes to organic chemists. Special attention is given certain selectivities which can be obtained using coordination complexes as catalysts. Attention is also given related polymer supported complexes. For research collections on organic chemistry syntheses. The author is with University of Sussex (U.K.). (RGK)

*Handbook of heat and mass transfer. Volume 1: Heat transfer operations*. Edited by Nicholas P. Cheremisinoff. Houston, TX: Gulf Publishing; 1986. 1456 pp. $165. ISBN 0-87201-411-8. *Volume 2: Mass transfer and reactor design*. 1518 pp. $165. ISBN 0-87201-412-6.

This two-volume set is aimed at unifying heat and mass transport concepts for the practicing engineer. Provides fundamental understanding of the physical processes and detailed guidance in applying principles to designing complex unit operations and chemical reactors involving multiphase flows. Volume 1 concerns heat transfer mechanisms and industrial operations and design. The topics in Volume 2 are mass transfer principles, distillation and extraction, multiphase reactor systems, and special applications and reactor discussions. A must for collections on chemical engineering and physical chemistry. The editor is with Exxon Chemical Co. (RGK)

*Oxazoles*. Edited by I. J. Turchi. New York: Wiley; 1986. 1064 pp. $225. ISBN 0-471-86958-9. (The chemistry of heterocyclic compounds. v. 45.)

This massive volume continues the extensive series of monographs on the chemistry of heterocyclic compounds. It represents a survey of the published work on the synthesis, reactions, and spectroscopy of mononuclear oxazoles, oxazolones, and mesoionic oxazoles. Very heavily documented (over 2,400 citations). For extensive organic chemistry reference collections in industry and academe. (RGK)

*Practical protein chemistry—a handbook*. Edited by A. Darbre. New York: Wiley; 1986. 620 pp. $95. ISBN 0-471-90673-5.

This is essentially a collection of techniques now available to the protein chemist, a timely contribution because the impact of generic engineering and protein engineering has made on the value of such scientists. The work's principal value is for the average biochemist who wants to practice protein chemistry but has been discouraged by the need for a single *vade mecum*. Very heavily documented and very well-indexed. The editor, and contributor of the book's most extensive chapter, is with King's College, London. (RGK)

*Propane, butane, and 2-methylpropane*. Edited by Walter Hayduk. New York: Pergamon; 1986. 445 pp. $100.00. ISBN 0-08-029202-X. (Solubility data series v. 24.)

This volume contains the result of a comprehensive search of the literature (to the end of 1983) on the solubility of three gases in liquids. Propane, butane, and 2-methylpropane are unusual gases and their rather unusual solubility characteristics are depicted in these 429 pages of tables. Included mainly are critical evaluations, experimental values, analytical procedures, and literature references. There are three indexes: by system, by registry number, and by author. For all comprehensive chemistry collections in academe and industry. (RGK)

*(The) properties of gallium arsenide*. Edited by J. C. Brice. New York: INSPEC; 1986. No pagination. $195.00. ISBN 0-85296-323-8. (EMIS. Data reviews series no. 2.)

This is a handbook of evaluated numeric data and reviewed knowledge distilled by those working at the cutting edge of gallium arsenide research. Provided are physical, electronic, and optical properties, carrier attributes, various spectra, and surface structure and oxidation, plus numerous other categories of data. Well-indexed. For electronic engineers, material scientists, and physicists. Note too that this work has been produced from the EMIS databank which is available for updates as file 105 on the European Space Agency Information Retrieval System. (RGK)

*Pyridine-metal complexes*. By Piotr Tomasik and Zbigniew Ratajewicz. 3-volume set. New York: Wiley; 1985. 2247 pp. $595.00 (The chemistry of heterocyclic compounds. v. 14, parts 6A, 6B, 6C.) ISBN 0-471-05073-3.

This is the first comprehensive review of the coordination compounds of pyridine, pyridine N-oxide, and their ring-substituted derivatives with known metals capable of forming such complexes. Topics covered are: theoretical overview, preparative methods, structural aspects, physical methods of analysis, as well as chemical and physicochemical properties, and biological activity. A whole spectrum of applications is noted. Complexes are compiled in tables according to the metal providing direct access to the key references. Eleven thousand literature references are included. For all organic chemistry collections in academe and industry. No subject index. (RGK)

*(The) pyrimidines. Supplement II*. By D. J. Brown. New York: Wiley; 1985. 916 pp. $195.00. (The chemistry of heterocyclic compounds. v. 16.) ISBN 0-471-38116-0 (v. 1).

This supplement stands as a review of the pyrimidine literature for the period 1968-1983, inclusive. It is important to note that *Supplement II* does not contain any information already recorded in the original volume (covering all literature through 1957) and *Supplement I* (for 1958–1967). Thus all three volumes must be used together in order to cover all the literature on any aspect of pyrimidine chemistry. Reference prior to no. 4329 will be found in one or the other of the earlier volumes. Patents have been ignored in general. The subject index is extensive. For all organic chemistry collections in academe and industry. (RGK)

*Reflexions on the motive power of fire*. By Sadi Carnot. Translated and edited by Robert Fox. New York: Lilian Barber Press; 1986. 230p. $39.50. ISBN 0-936508-16-7.

In this short, brilliant, and classic work, *Reflexions sur la puissance matrice du feu* (1824), Sadi Carnot lays some of the main foundations of classical thermodynamics, e.g., what we know now as the second law of thermodynamics. Also included in the volume are translations of Carnot's other scientific manuscripts. This is a very critical (but not necessarily negative) English edition of all the works. The commentaries are extensive and of course provide numerous corrections and calculations. This is an important contribution which belongs in all history of science and technology collections. The translator is with the University of Lancaster, U.K. (RGK)

*Thermochemical data of organic compounds*. 2d ed. By J. B. Pedley and others. New York: Chapman and Hall; 1986. 791 pp. $120. ISBN 0-412-027100-1.

The purpose of this massive collection of material is to enable users of thermochemical data to predict values of standard enthalpies of reactions involving organic compounds ranging in complexity from simple alkanes to biologically important compounds such as amino acids. Most of the data is developed for some 3,000 organic compounds of the elements C, H, O, N, S and the halogens. Does not contain data on radicals, ions, organometallic compounds or multicomponent systems. Properties other than standard enthalpy of formation at 298.15K are not included in this edition. For all organic chemistry collections in academe and industry. (RGK)

## SCIENCE, GENERAL

*Handbook of laboratory waste disposal*. By Martin J. Pitt and Eva Pitt. New York: Wiley; 1985. 360 pp. $87.95. ISBN 0-470-20202-5.

This book is about all types of laboratory waste and their practical management. Provides sensible advice and information culled from a wide range of sources. The authors pound on the theme that waste *can* and *should* be managed. Methodology and techniques form the bulk of the text. The appendices include an abbreviations glossary, critical listings of commercial products and supplies, useful chemical tables including USA Hazardous Waste and UK Special Waste specifications, and names of those chemicals which can deteriorate and the resulting waste disposal hazard. Some illustrative matter. For all collections dealing with medicine, biology, chemistry, technology, and engineering (at the very least). First author with Loughborough University of Technology; second with University of Aston, Birmingham. (RGK)

*Maritime affairs—a world handbook*. By Henry W. Degenhardt. Detroit: Gale Research; 1985. 412 pp. $90.00. ISBN 0-8103-2051-7.

Provides coverage of subjects relating to oceans and seas, including international maritime law, sea transport and communications, the exploitation and conservation of marine natural resources, scientific research, boundary disputes, and the military dimension of the sea. There are dictionary sections giving descriptions, names, and addresses for major international maritime organizations, major organizations for each subject, maritime publications arranged by subject and country, the full text of the U.N. Law of the Sea Convention, and a subject-arranged bibliography of about 2,000 references. For academic, government, and the larger public libraries. (RGK)

*(The) solid waste handbook: a practical guide*. Edited by William D. Robinson. New York: Wiley; 1986. 811 pp. $74.95. ISBN 0-471-87711-5.

This practical guide presents a rather comprehensive single source reference for current solid waste management public issues, implementation issues, and hazardous waste administration guidelines. It examines areas such as legislation, finance, technologies, and assessments for the future. It is definitely *not* a design engineering manual. The work is for technical, legal, and finance professionals and is a guide for all levels of government. Well-documented. (RGK)

## EXTANT

*Fieser and Fieser's reagents for organic synthesis*. Vol. 12. By Mary Fieser. New York: Wiley; 1986. 643 pp. $47.50. ISBN 0-471-83469-6 (v. 12).

This volume continues the famous reagents series and for the most part includes references to papers published during 1983 and 1984. There is the usual index of reagents according to types (e.g., chlorination) plus subject and author indexes. An absolute must for every collection on organic chemistry synthesis. (RGK)

# *SCI-TECH COLLECTIONS*

Tony Stankus, Editor

Colette O'Connell has prepared a detailed paper on computer-aided design and computer-aided manufacturing (CAD/CAM), a topic of great importance. Because of the length of her paper it was necessary to publish only Part I in this issue, covering the history and literature of the subject. In the following issue the concluding part will be published. It will contain a glossary, a classified bibliography and other sections.

All those considering preparing a paper in which the literature of some topic is analyzed are once again invited to contact this editor regarding the possibilities of having such a paper published in this journal. Your inquiries will be welcome.

 

# CAD/CAM (Computer-Aided Design/ Computer-Aided Manufacturing): A History of the Technology and Guide to the Literature Part I

Colette O'Connell

## *HISTORY OF THE TECHNOLOGY*

### *Introduction*

In 1947 the Analysis Laboratory at the California Institute of Technology introduced analog computers as a design aid in the aircraft industry.[1] The use of computers to solve engineering design and manufacturing problems increased over the next four decades. In the 1980s Computer-Aided Design/Computer-Aided Manufacturing (CAD/CAM) has emerged as the dominant technology for engineering design and manufacturing. This paper outlines CAD/CAM's development and its literature. Since a list of CAD/CAM related acronyms will be useful, a glossary of acronyms is included in *Appendix 1*. *Appendix 2* contains a classified bibliography for a CAD/CAM library collection. Directory information for pertinent organizations and research centers is listed in *Appendix 3*.*

Colette O'Connell has a BA degree in Mathematics from Rhode Island College in Providence, RI, and a MLS from the State University of New York at Albany. She is the Engineering Librarian and Coordinator of Online Search Services at Folsom Library, Rensselaer Polytechnic Institute, Troy, NY 12181.

*See Volume 8, Number 1 for Part II of this article which includes Appendices 1, 2, and 3.

 

## *Applications Programs and Time Sharing*

Computer-Aided Engineering (CAE) involves the application of computers to assist the engineering process. CAE faced two problems in the 1950s. First, engineer-computer communication was hampered by the lack of engineering applications computer languages. About 1956 Automatically Programmed Tools (APT) appeared as the first computer language with a specialized engineering application.[2] APT was developed at the Massachusetts Institute of Technology (MIT) through United States Air Force sponsored research. A number of companies were also involved in its development.

Around the same time the FORTRAN (FORmula TRANslation) computer language was introduced. FORTRAN allowed for the expression of problems in algebraic terms. FORTRAN was widely used and provided engineers and scientists with an easier means of communicating with computers.

Developed in 1961 AUTOmatic PROgramming of Machine Tools (AUTOPROMT) was a simulated prototype command language. Based on APT, AUTOPROMT was a cooperative effort in which IBM, United Aircraft Corporation, and Pratt & Whitney Co., Inc. were the principal collaborators. With AUTOPROMT a computer could produce three-axes machine tool control tapes by describing the surface to be generated rather than the traditional method of plotting the path of the tool point. AUTOPROMT translated a blueprint into a finished part. "The AUTOPROMT system was viewed by engineers as a step toward highly sophisticated machine manufacturing."[3] Other applications oriented computer languages followed. COordinated GeOmetry (COGO) and STRuctural Engineering System Solver (STRESS), both products of the Massachusetts Institute of Technology (MIT), Civil Engineering Department are examples.

Computer availability posed a second problem. In the 1950s batch processing was the standard means of computer processing. Batch processing was frustrating because its turnaround time was slow. The solution was time sharing. Systems programs provided time sharing whereby a central computer serviced a large number of remote terminals. The appearance is that each user has full use of the computer facilities when in fact he/she is one of many simultaneous users. Although systems programs offered a solution, long distance communications was still in the early stages of development. About this time, both MIT and the Griffiss Air Force Base at Rome, NY successfully established time sharing facilities where remote terminals were attached to a large centralized computer.[4]

### *Early Research and Development*

Applications programs and time sharing facilitated the development of interactive computer systems in the 1960s. Most systems at that time resulted from corporate research and development programs in companies such as General Dynamics Corporation and Boeing Company who could afford the development costs. By 1961 General Dynamics Corporation's Electronics Division produced computer-developed designs which were drawn electronically.[5] In 1964 Boeing used cathode-ray tube (CRT) devices to produce drawings for the Dyna-Soar space vehicles.[6] By 1966 computer-aided design systems were in use at the General Motors Technology Center, IBM's Kingston NY facility, Lockheed-Georgia Company, and the Boeing-Seattle Aerospace Division.[7]

In the university community MIT played a pivotal role in the development of Computer-Aided Design (CAD). In the early 1950s MIT's Servomechanism Laboratory developed the first automatically controlled milling machine.[8] At this time computers were seen as "a new force on the horizon."[9] MIT's SKETCHPAD Project is a benchmark for CAD. SKETCHPAD, a robot draftsman was developed in the early 1960s by Dr. Ivan Sutherland of MIT.[10] The original SKETCHPAD program was a simple circuit diagram produced on a TX-2 computer scope at MIT's Lincoln Laboratory. Its principal components were a display scope (CRT) and a light pencil. Complicated drawing capabilities were available, but not yet economical.[11] At the time MIT was initiating the SKETCHPAD system a similar program which began in 1959 was underway at General Motors.[12]

In 1964 IBM 1627 and 7404 computers were used to produce maps and contour plots, as well as space missile graphs and charts.[13] The same year a Princeton professor developed an electronic map plotter for use by building contractors.[14] In 1966 T. E. Johnson of MIT developed SKETCHPAD III which also used the Lincoln TX-2 computer. SKETCHPAD III allowed the study of three-dimensional space linkages.

In 1965 the National Aeronautics and Space Administration

(NASA) introduced an automated drafting system. This system called NASA STructural ANalysis (NASTRAN) was developed over a three year period and involved the cooperative efforts of the Computerservices Corporation, Martin-Marietta Baltimore Division, and MacNeal-Schwendler Corporation under the direction of the Goddard Space Flight Center.[15] NASTRAN allowed a designer to sketch free hand in one quarter of the time required for conventional drafting techniques. In 1972 the first annual NASTRAN Users' Colloquium was held.

### *Applications*

Throughout the 1960s new computer-aided design (CAD) and computer-aided design and drafting (CADD) systems appeared. Often the names CAD and CADD were used interchangeably. Initially CAD systems were used primarily for architectural design. One of the first engineering industries to use CAD on a large scale was the electronics industry. CAD easily met the increasing demands for circuit designs. The oil and chemical companies and civil engineering firms followed using CAD for pipe layouts and construction projects. Mechanical design and manufacturing applications are the biggest growth areas since 1970,[16] and mark the emergence of CAD/CAM.

### *Computer Graphics Technology*

By 1967 computer graphics engineering applications were well established and included equation generation, iterative design, design verification, and simulation. Computer-assisted engineering graphics was one of the major technological developments of the decade. Computer graphics accommodated both design and manufacturing activities and was the key to CAD/CAM's development.

Interactive computer graphics systems required little knowledge to operate. Graphics technology included drafting and digitizing where computers translated from digital data to drawings and from drawings to digital data. Output devices included printers, X-Y plotters, drafting machines and CRT displays.

By 1978 more types of input and output devices were available, and consisted largely of keyboards, pointing devices and tablets. Display devices included direct view storage tubes, stroke or vector writing units, raster scans, scan convertors and plasma panels.

Other improvements included raster scan CRT displays, more intelligent graphics displays, microprocessors, and larger memory.

The programming of two-dimensional and three-dimensional displays was still a problem. Their manipulation had been mastered, but the need to develop algorithms which used less computer time still existed. Quality color graphics, indefinite screen display, and ease in screen modification were also needed.[17]

## *Computer Graphics Standards*

Computer graphics standards have been an ongoing concern. The Association for Computing Machinery Special Interest Group on Computer Graphics (ACM/SIGGRAPH) held the Workshop on Machine Independent Graphics in 1974 which was the beginning of formal graphics standards deliberations.[18] The workshop resulted in the establishment of the SIGGRAPH Graphics Standards Planning Committee (GSPC). In May 1976 the Workshop on Graphics Standards Methodology was held in Seillac, France. At this meeting GSPC and others established two objectives: (1) portable graphics standards and (2) division between modelling and viewing of graphics objects.[19] By the late 1970s graphics standards committees were at work in the United States.

## *Specific Standards*

Common Operational Research Equipment (CORE) was a standard developed by the SIGGRAPH GSPC. The first draft of CORE was presented at the SIGGRAPH '77 Conference, and the second draft was presented at SIGGRAPH '79. Many software programs have been developed for CORE. CORE permits three-dimensional as well as two-dimensional graphics. The American National Standards Institute's (ANSI's) failure to adopt CORE undoubtedly contributed to CORE's replacement by Graphics Kernel System (GKS).[20]

GKS was developed by Deutsches Institut fuer Normung (DIN), the German Standards Institute. GKS is a standard for a graphics programming interface, a standard computer graphics subroutine package. GKS has an advantage over the other graphics standards in that an application written for GKS will usually run on another system without modification.[21] GKS was the first international standard to reach draft proposal stage in 1979.

Presently GKS has a two-dimensional capability, but a three-dimensional capability is under development.[22] Both ANSI and the International Organization for Standardization (ISO) support GKS. Programmer's Hierarchical Interactive Graphics Standards (PHIGS), a similar standard under consideration, offers three-dimensional computer graphics.[23]

North American Presentation Level Protocol Syntax (NAPLPS) defines graphics-devices software interfaces. NAPLPS was developed by a committee of the Canadian Standards Association. ANSI adopted NAPLPS in 1983.

Initial Graphics Exchange Specification (IGES) was the result of collaboration between the major aerospace companies and the United States National Bureau of Standards (NBS). IGES was first published in January 1980. Version 1.0 is part of ANSI Standard Y14.26M (*Computer-Aided Preparation of Product Definition Data*) published in 1981. IGES offers a graphics data interchange between CAD/CAM systems from different manufacturers. Version 2.0 of IGES was approved in July 1982, and is available from the National Technical Information Service/National Bureau of Standards as PB83-137-448-2.0. IGES may be a way to achieve CAD/CAM translation programs, but does not yet have unanimous backing.[24]

AN ISO subcommittee has proposed a universal standard for data exchange from different CAD/CAM systems.[25] The standard, STandard for Exchange of Product data (STEP) will replace the American IGES and its German and French counterparts. Although a draft is expected by the end of 1986, widespread use is not anticipated until the 1990s.

ANSI is working in cooperation with European standards groups such as ISO.[26] Future development of microcomputer graphics will benefit substantially from ANSI and ISO cooperation, but require special consideration before achieving interactive interfaces.[27] Standards for personal computer software should make machine-independent applications software possible.

General Motors Corporation has established Manufacturing Automation Protocol (MAP) which selects existing or emerging standards for multi-media communication in the manufacturing setting.[28] If a needed standard is not available, General Motors develops an interim standard. GM's size and purchasing power has been an incentive for vendor compliance. GM's MAP is especially important because its objective is integration of CAD/CAM systems on a much larger scale than the standards discussed previously.

### *Later Corporate Development*

McDonnell Aircraft Company developed Computer-Aided Design-Drafting (CADD) in 1970 to support its F-15 project.[29] The Lockheed California Company of Burbank marked a milestone in 1974 with the release of Computer-Aided Design and Manufacture (CADAM), an off-the-shelf software system.[30] CADAM was the result of ten years research and development, and cost $3 million. CADAM decreased production time five fold, was ten times faster than earlier systems, and offered medical as well as engineering applications. Today Lockheed's CADAM is one of the most widely used CAD/CAM systems.[31]

In 1977 Martin Marietta Aluminum of Torrance, California introduced a mini-computer CAD/CAM system into its Tooling Division.[32] In addition to machining activities, the Marietta mini-computer system provided cost data to the sales department. Produced by Computervision Corporation, the new system cost $225,000 with an annual operating expense between $20,000 and $25,000. This system, which was user-friendly, required six to seven weeks training, and promised user proficiency in six months.

### *Numerical Control*

MIT in cooperation with the U.S. Air Force pioneered the use of numerical control (NC).[33] NC refers to the use of computer graphics to control and regulate manufacturing tools such as lathes and die cutters. A numerical control system is a CAD/CAM system designed to automate a single manufacturing process. Direct numerical control (DNC) uses one computer which feeds data to a group of NC machines as needed. A DNC includes the hardware and software required to drive one or more NC machines simultaneously while connected to a common computer memory. The first DNC system began operation in 1968.[34]

The movement in the late 1970s from direct numerical control (DNC) to computer numerical control (CNC) was a breakthrough for CAD/CAM. The CNC system uses a dedicated microcomputer within the numerical control device that allows data to be input directly to the NC device. A CNC system brings the computer into more direct contact with the machine tool, and is often linked directly to the central computer. The first CNC appeared about 1970.[35]

Numerical control is one of the oldest automated manufacturing activities. Until recently CAD/CAM principally involved product layout and aids to the manufacturing process,[36] such as numerical control. Now CAD/CAM functions also include activities such as robotics and factory management. Solid modelling is a major new component over the past two or three years.

## *CAD/CAM*

The organization and management of manufacturing as a process industry was a revolutionary idea in the early 1970s. The seventies were hailed as the Second Industrial Revolution with CAD/CAM as the key element.[37] February 1972 marked the first CAD/CAM Conference held by the Society of Manufacturing Engineers.

In 1973 IBM viewed the future for CAM in the following areas: (1) machine monitoring, (2) machine control, (3) facilities and environmental testing, and (4) mechanical and electrical testing.[38] The major concern then, as it is now, was the integration of design and manufacturing, that is, how will it be integrated and who will be responsible for its integration.[39] A number of factors limited CAD/CAM use in the late seventies. High cost was a major barrier. Many CAD/CAM systems cost well over $100,000.[40] Mainframe costs were dropping in 1978, and microcomputers were not yet sophisticated enough to handle CAD/CAM systems.

As microcomputer based systems gained the capability to support CRT terminals, the cost per terminal dropped. While hardware costs were dropping, software development was still time consuming and expensive. A lack of understanding of computer-aided manufacturing still existed.[41] Attempts to quantify productivity were difficult for management to justify installation of CAD/CAM systems, especially for smaller companies.

Today CAD/CAM is becoming more affordable for smaller companies. Most turnkey systems which make up the greatest proportion of CAD/CAM systems use microcomputers.[42] Personal computers had been too slow, but the possibility of upgraded microcomputers has helped users to adopt personal computers for CAD/CAM applications. Some desktop computers can be configured for computer graphics. The possibility of putting the most commonly used graphics routines on one chip should further facilitate CAD/CAM's expansion.[43]

New graphics devices have replaced keyboards as the primary

means of entering data. Touch screens, track balls, and mice have joined light pens and joy sticks as input devices. New third party software seems to appear weekly, and will continue to be developed because more potential CAD/CAM applications exist than vendors to develop software.[44]

General Electric was probably the biggest single corporate user of interactive computer graphics in the early 1980s.[45] In 1985 McDonnell Aircraft Company, Deere and Company, Westinghouse Defense and Electronics Center, General Motors Corporation, and Ingersoll Milling Machine Company were among the leaders in CAD/CAM use.[46]

CAD/CAM has had three stages of growth: (1) single applications systems with one or two terminals, (2) multiple terminals for single applications, and multiple applications on a single terminal, and (3) integration of systems and functions.[47] Social, economic, and technical factors have contributed to CAD/CAM's growth. Increased preference for service industry as opposed to manufacturing industry employment has decreased the supply of industrial workers.[48] Industry has had to rely more on automated design and manufacturing technologies. The government is increasing requirements for a safe workplace, and employers are providing safer work environments. As a result a number of manufacturing activities have been automated. The rising cost of manufacturing labor relative to manufacturing productivity has forced companies to turn to CAD/CAM. In addition CAD/CAM's benefits include increased productivity and analytical capabilities, enhanced user creativity and reduced product and development costs.[49]

Manufacturing activities follow three stages which make it suitable for computers: (1) collection of information, (2) decision making, and (3) dissemination of orders.[50] Although manufacturing accounts for about thirty percent of the gross national product,[51] it has not been a highly productive industry. CAD/CAM is seen as the key to the manufacturing industry's survival.[52]

CAD/CAM is emerging as a key technology to increase productivity as well as improve quality. While the computer industry is growing at a rate of six percent annually, the CAD/CAM industry's annual growth has been thirty to forty percent in the early 1980s.[53] In 1984 CAD/CAM industry growth rate reached fifty percent, and systems sales equalled $2.9 billion.[54] The seven industry leaders (IBM, Computervision, Intergraph, CALMA, Applicon, Prime and McAuto) accounted for eighty percent of the sales. IBM is expected to become the industry leader in the next year. American companies spent $930 million on automated design equipment in 1983 and are expected to increase this amount to $2.8 billion in 1987 and $9.8 billion in 1995.[55]

### *Cooperative Programs*

Several cooperative programs were established in the 1970s. Computer-Aided Manufacturing-International (CAM-I) is a not-for-profit research and development cooperation for the advancement of computer applications to manufacturing problems. Founded in 1972 by groups from industry, academia, and government, its membership is approximately 160 and includes Boeing, Mitsubishi Electric Corporation, MIT, Rensselaer Polytechnic Institute (RPI), and the University of Manchester (UK). CAM-I sponsors cooperative research and development, and conferences.

Its publication program includes conference proceedings, reports, software, and reference manuals. CAM-I originated Computer-Aided Process Planning (CAPP) developed by McDonnell Douglas Automation Company as a prototype computer program to produce manufacturing process plans.[56]

Integrated Computer-Aided Manufacturing (ICAM) is a United States Air Force program which provided "seed money" to advance the manufacturing technology frontier. Established in 1978 ICAM initially was funded by the United States Department of Defense and managed by the United States Air Force Materials Laboratory (Wright Patterson Air Force Base, Ohio). Its goal has been to develop a program to coordinate computer-aided design and computer-aided manufacturing functions now used "piecemeal" by industry.[57]

Today ICAM is the cooperative effort of European countries and the Air Force. Its research and development programs include the development of manufacturing software, group technology and cellular manufacturing. Established in 1976 with a nine year budget, ICAM will culminate with a demonstration integrated sheetmetal center at the Boeing Military Airplane Company.[58]

ICAM has worked closely with another government CAD/CAM program, Integrated Program for Aerospace-Vehicle Development (IPAD). IPAD was a NASA sponsored project which began in 1976. IPAD was also a computer program developed by Boeing Company under contract with NASA. The IPAD computer program was a software package for integration of CAD functions. The

IPAD organization is now concerned with database management and integration for CAD/CAM systems.[59]

The Automated Manufacturing Research Facility (ARMF) of the United States National Bureau of Standards has a number of CAD/CAM programs underway. The ARMF serves as a test center for computer-controlled machinery which will be used by industry.[60]

### *Computer-Integrated Manufacturing*

Computer-integrated manufacturing (CIM) is the integration of all aspects of manufacturing by means of computer technology. It includes four sectors: (1) factory automation; (2) manufacturing control; (3) manufacturing planning; and (4) engineering design.[61] CIM has also been defined as a manufacturing enterprise in which (1) all functions can be expressed as data; (2) data can be manipulated by computers; and (3) data moves between manufacturing activities throughout the life of a product.[62] CIM consists of software and hardware for product design, production, planning, and production processes. To date Deere and Company in Waterloo, Iowa is the best example of a CIM plant.[63]

CAD/CAM is but one component of CIM. CIM encompasses related technologies such as flexible manufacturing systems (FMS), robotics, management information systems (MIS), and group technology. Group technology is a system where parts are organized into families based on similarities. Standard plans for a part are stored in a computer and customized plans for new parts are developed by editing the standard plans.

### *System Integration*

"Islands of automation"[64] are CAD/CAM's biggest problem. Vendors have developed CAD/CAM systems independently, and companies have implemented them in accordance with their individual management styles. Standardization between CAD and CAM is difficult. The basic problem is interfacing, "transfering databases between areas such as engineering design and manufacturing."[65] In general CAD/CAM systems are not integrated, which leads to a proliferation of databases.

Entry of existing data into a CAD/CAM system is another unresolved problem. Most manufacturing data is geometric, but an increasing amount such as materials properties is nongeometric.

Database management systems must accommodate various types of data. Issues such as data communication, consistency, and integration must be resolved. Organizational issues to be addressed include fragmentation and isolation of departments, top management support, responsibility for implementation, and labor supply.[66]

A CAD/CAM fully integrated product can be designed and built from start to finish through interactive computer graphics.[67] Today a typical integrated system includes most if not all of the following: geometric modelling, automated design/drafting, finite element modelling and analysis, and numerical control.[68]

The key to an integrated system is a common engineering and manufacturing database.[69] Exchange of data between systems is important. The major automotive manufacturers now require that their CAD/CAM vendors can communicate electronically with them.[70] Another step in the ''marriage'' of CAD/CAM is its link to other technologies such as robotics and expert systems. The question remains who will integrate CAD/CAM systems: the vendors or customers. Full system integration may not arrive until 1988 at the earliest.[71] GM's MAP will undoubtedly have a major role in full system integration.

The use of personal computers with CAD/CAM systems with single stations or networked operations is more widespread. Existing drawings are being digitized for use in CAD/CAM systems. More emphasis has been placed on training for CAD/CAM operators. CAD/CAM curricula are more prevalent. Establishment of the College CAD/CAM Consortium,[72] a cooperative program of eighteen leading U.S. engineering schools, is an indication of university commitment to CAD/CAM. Funded by the National Science Foundation, the College CAD/CAM Consortium aims to develop educational materials for the field.

Database management systems are more widely used. The appearance of CAD/CAM service bureaus in 1984 is another indication of CAD/CAM's growth. Some other recent developments include three-dimensional CAD/CAM systems, full factory automation, and distributed workstations. Voice data entry as an input device will be routinely available by the year 2000.[73] The emergence of expert systems is exciting because these systems take advantage of existing knowledge.[74] Expert systems are computer systems with high levels of performance in areas which for humans would require years of specialized education and training.[75]

CAD/CAM systems have not been evaluated yet. A study at

MIT's Productivity Research Program is in its third year of evaluating the effectiveness of CAD/CAM systems. The National Research Council/Manufacturing Studies Board (NRC/MSB) has investigated current industrial efforts to improve CAD/CAM interfaces. NRC/MSB has also made recommendations to advance United States CIM technology, and formed a committee on CAD/CAM interfaces. The committee has studied five companies and three major government programs that are leaders in CAD/CAM integration. NRC/MSB committee recommendations are: (1) NASA should adopt CIM, (2) companies should form CIM consortia, (3) the Computer and Automated Systems Association of the SME should serve as a centralized CIM information center, (4) federal research and development on CIM should continue, (5) federal agencies should accept IGES, and (6) manufacturers should consider adoption of CIM systems.[76] It is too soon to evaluate the impact of the NRC/MSB recommendations. SME has established a CAD/CAM information center, CARIC (Computerized Automation and Robotics Information Center) as well as an online information retrieval service, *INTIME Manufacturing DataBank*. The development of STEP and General Motors' MAP will undoubtedly affect recommendation five.

## *THE LITERATURE*

### *Dictionaries, Handbooks, and Directories*

Computer, electrical, electronics and manufacturing engineering ready reference books usually cover CAD/CAM. However, some dictionaries, encyclopedias, handbooks and buyers' guides are available specifically for this field. *Glossary of Computer-Aided Manufacturing*, *Handbook of Design Automation*, and *CAD/CAM Handbook* are examples. The Marquis' *Who's Who of Computer Graphics* lists electrical and mechanical CAD/CAM graphics professionals as well as architectural and engineering CAD professionals.

In addition to name and address information most directories for the field are buyers' guides. These directories also include product reviews and evaluation guidelines for prospective buyers. Some directories include guides to the literature, association listings, and conference announcements. *Computer Graphics Marketplace* and *Robotics, CAD/CAM Market Place* are buyers' guides which include this type of information. Some trade periodicals provide buyers' guides with their subscriptions. *Computer-Aided Engineering* and *Machine Design,* both Penton-IPC publications, have special issues that are product directories.

### *Bibliographies, Indexes, Abstracts*

Several CAD/CAM bibliographies are available. Engineering Information, Inc. produces *Technical Bulletin Series* which includes a CAD/CAM and a computer graphics bibliography. Bibliographies in the *Technical Bulletin Series* consist of references from the Engineering Information database, *COMPENDEX*. NTIS produces CAD/CAM bibliographies from the *NTIS* and *COMPENDEX* databases. The Library of Congress Science and Technology Division Reference Section has published *CAD/CAM (Computer-Aided Design/Computer-Aided Manufacturing): A Brief Guide to Materials in the Library of Congress* (LC Tracer Bulletin TB85-7). Vance Bibliographies has published *Computer-Assisted Design/Computer-Assisted Manufacturing: A Select Bibliography.*

Indexing tools covering engineering and computer science are the best CAD/CAM information sources, and many have online counterparts. *Applied Science and Technology Index, Engineering Index Monthly, Ei Engineering Conference Index*, and *Science Abstracts* offer excellent coverage for this field. All are available online. *Science Citation Index*, available in printed and online formats, is also a good source. *The Computer Database*, DIALOG's File 275, is a very good information source which does not have a printed counterpart.

EIC/Intelligence Publishing Division produces *CAD/CAM Abstracts*. In addition to abstracts this publication lists technical reports, patent references and corporate activities, covering both the technical and business areas. *CAD/CAM Abstracts* is also available as part of DIALOG's File 238, *SUPERTECH*. Additional coverage of the business aspects is offered by *Business Periodicals Index* and *PTS F&S Index*. Both are available online.

Printed indexes offered by organizations such as the Association for Computing Machinery (ACM), the Institute of Electrical and Electronics Engineers (IEEE) and the Society of Manufacturing Engineers (SME) are good sources for CAD/CAM information, but they usually do not have commercially available online formats. The Society of Manufacturing Engineers offers *INTIME Manufac-*

*turing DataBank*, a database of SME's technical papers, books, and journal articles. *INTIME* is probably one of the most important databases for the CAD/CAM field. Although *INTIME* is not a commercially available database, its search services as well as published bibliographies are available to the public for modest fees.

### Online Database Survey

Five technical databases (*COMPENDEX*, *Ei MEETINGS*, *INSPEC*, *NTIS*, and *The Computer Database*) were searched on DIALOG in April 1986. The search was confined to a five-year period, 1981 to 1985. The following strategies were used:

1. CAD/TI,DE,ID or COMPUTER()AIDED()DESIGN/TI,DE, ID
2. CAD/CAM/TI,DE,ID,
3. COMPUTER()AIDED()MANUFACTUR?/TI,DE,ID, and
4. COMPUTER()INTEGRATED()MANUFACTUR?/TI,DE,ID.

The subject descriptors were qualified to the title, descriptor and identifier fields. Since the databases vary in size, search results are presented in percentage of the total database for the time period searched. Although The Computer Database only covers 1982 to present and does not have an identifier field, it was included because of its subject coverage. Figures 1 and 2 summarize the search results.

### Technical Reports

*Government Reports Announcement and Index* and *Scientific and Technical Aerospace Reports* are excellent sources for identifying technical reports. The Department of Energy, the National Aeronautics and Space Administration, and Wright-Patterson Air Force Base all have technical report series which include CAD/CAM publications. The CAD/CAM research centers at various engineering schools frequently publish technical reports as part of their publication programs. Universities with CAD/CAM research facilities are listed in *Research Centers Directory*. Directory entries indicate if a center has a publication program.

### Conference Proceedings

Most conferences in the field are sponsored by the major CAD/CAM professional organizations: the Association for Comput-

FIGURE 1

SUBJECT SEARCH - TI,ID,DE

(1981-1985, inclusive)

(percentage)

CAD/TI,ID,DE or COMPUTER()AIDED()DESIGN/TI,ID,DE

| COMPENDEX | Ei Meetings | INSPEC | NTIS | The Computer Database* |
|---|---|---|---|---|
| 4.55 | 6.85 | 5.63 | 2.797 | 6.03 |

COMPUTER()AIDED()MANUFACTUR?/TI,ID,DE

| COMPENDEX | Ei Meetings | INSPEC | NTIS | The Computer Database* |
|---|---|---|---|---|
| 1.01 | 3.56 | 0.147 | 0.885 | 0.407 |

CAD()CAM/TI,ID,DE

| COMPENDEX | Ei Meetings | INSPEC | NTIS | The Computer Database* |
|---|---|---|---|---|
| 0.170 | 0.830 | 0.814 | 0.080 | 1.97 |

COMPUTER()INTEGRATED()MANUFACTUR?/TI,ID,DE

| COMPENDEX | Ei Meetings | INSPEC | NTIS | The Computer Database* |
|---|---|---|---|---|
| 0.062 | 0.58 | 0.174 | 0.011 | 0.030 |

*limited to 1982-1985 and TI,DE

ing Machinery (ACM), the Institute for Electrical and Electronics Engineers (IEEE) Computer Society, the International Federation for Information Processing (IFIP), the Society of Manufacturing Engineers (SME), and the Computer and Automated Systems Association of the SME (CASA/SME). The Design Automation Conference is an annual conference which was first held in 1964. Design Automation

Conference bills itself as "the premier conference dealing with the subjects of computer-aided engineering, computer-aided design, and computer-aided manufacturing."[77] The conference proceedings are a joint publication of the Association for Computing Machinery and the Institute of Electrical and Electronics Engineers.

A number of CAD/CAM conferences have been sponsored by the IFIP Technical Committee 5 on Computer Applications in Engineering (IFIP TC-5). SME and CASA also publish a number of CAD/CAM related conferences. ASSEMBLEX Conference, CAD/CAM Conference, and the AUTOFACT Conference are included among these. The North American Manufacturing Research Conference is also published by SME. SIGGRAPH Conference is an annual ACM conference for computer graphics. In addition to society sponsored conferences, many commercially published conference proceedings are available. These proceedings are often one-time publications.

### *Trade Publications, Research Journals, and Newsletters*

Some journals focus on CAD/CAM. These include *Computer-Aided Engineering* (Penton/IPC), *CIM Technology* (CASA/SME) and *Computer-Aided Design* (Butterworth). The major societies,

FIGURE 2

SCISEARCH SUBJECT SEARCH RESULTS

| Publication Papers | Number of Papers | Number of Cited Papers | Number of Citing Papers* |
|---|---|---|---|
| 1977 | 5 | 5 | 1 |
| 1978 | 29 | 15 | 7 |
| 1979 | 41 | 74 | 18 |
| 1980 | 44 | 44 | 11 |
| Total | 119 | 138 | 37 |

*for a five year period subsequent to the publication date of the cited paper's publication date

IEEE Computer Society, ACM, SME and SME/CASA offer a mixture of basic periodicals and research journals. *Manufacturing Engineering* (SME) and *IEEE Transactions on Computer-Aided Design of Integrated Circuits* are examples.

*Design News* (Cahners) and *Machine Design* (Penton/IPC) are essential trade publications for any CAD/CAM collection. In addition to basic articles trade publications often include special buyers' guides, as well as software and patent announcements. Scholarly papers are included in commercially published journals such as *Computers and Industrial Engineering* (Pergamon) and *Computers in Industry* (Elsevier).

Several current awareness newsletters are available. These include *Manufacturing Technology Horizons* (Manufacturing Technology Press), *CIM Strategies* (Cahners), *Computer-Aided Design Report* (CAD/CAM Publishing) and *CAD/CAM Alert* (Management Roundtable). These newsletters generally offer a combination of business and technical information.

### *Citation Study*

A key word search was done on DIALOG's *SCISEARCH* (*Science Citation Index*) in April 1986. The search results were limited to English language journal articles. The following search strategy was used:

1. CAD/TI or COMPUTER()AIDED()DESIGN/TI or CAM/TI or COMPUTER()AIDED()MANUFACTUR?/TI or CAD() CAM/TI or COMPUTER()AIDED()MANUFACTUR?/TI or CAD/CAM/TI or CAM/TI
2. limit 1/eng
3. limit 2/art

The search covered the publication years 1977 to 1980, inclusive. CAD/CAM is a relatively new research area, so a multi-year subject search was required to yield a reasonable sample size. Two hundred references were retrieved, of which 119 had subject relevance and personal authors. A cited author search was done for each of these references. Cited author searches covered the five subsequent years to the cited author's publication date. For example a 1977 citation was searched as a cited reference for 1978 through 1982.

Thirty-seven of the 123 citations were cited a collective total of 138 times. The journals, *IEEE Transactions on Microwave Theory and Techniques* and *Computer-Aided Design*, occurred most often in the original search and were also among the most frequently cited journal titles. *IEEE Transactions on Microwave Theory and Technique* and *IEEE Transactions on Electron Devices* were the two journals citing the most papers.

If available, impact factors for journals from the original search, as well as for cited and citing journals, were collected from the 1984 *SCI Journal Citation Reports*.[78] *IEEE Transactions on Microwave Theory and Techniques* had one of the highest impact factors. Figures 3-5 summarize the citation study results and journal impact factors.

## *Monographs and Monographic Series*

*What every engineer should know about computer-aided design and computer-aided manufacturing* (1982) by Krouse, *Computer-aided design and manufacture* (1983) by Besant, and *Computer-*

FIGURE 3

Journal Titles from Original Subject Search

| Occurance | Impact Factor | Journal Title |
|---|---|---|
| 12 | 0.349 | Computer-Aid Design |
| 7 | 1.284 | Computer and Graphics |
| 5 | 0.000 | Manufacturing Engineering |
| 4 | - | Design News |
| 4 | - | Engineering |
| 4 | 0.174 | Journal of Mechanical Design: Transactions B |
| 4 | 0.042 | Telecommunications and Radio Engineering |
| 4 | - | Astronautics and Aeronautics |
| 3 | 0.132 | Electronics |
| 3 | - | Engineering Materials and Design |
| 3 | 1.211 | IEEE Transactions on Microwave Theory and Techniques |
| 3 | 0.016 | Toshiba Review |

FIGURE 4

Citing Journal Titles

| Frequency of Citing | Impact Factor | Journal Title |
|---|---|---|
| 24 | 1.211 | IEEE Transactions on Microwave Theory and Techniques |
| 19 | 2.342 | IEEE Transactions of Electron Devices |
| 15 | 0.342 | Computer-Aided Design |
| 12 | 1.004 | IEEE Journal of Solid State Physics |
| 12 | 0.355 | Journal of Optimization Theory and Applications |
| 7 | 1.132 | Electronics |
| 6 | 1.033 | IEEE Transactions on Circuits and Systems |
| 5 | 1.434 | Applied Optics |
| 5 | 0.950 | British Journal of Opthalmology |
| 5 | 0.422 | International Journal of Circuit Theory and Applications |
| 5 | - | Philips Technical Review |
| 4 | - | Nachrichlentechnische Zeitschrift |
| 3 | 0.821 | Automatica |
| 3 | 0.766 | IEEE Transactions on Systems, Man and Cybernetics |

*integrated manufacturing* (1979) by Harrington are basic texts which offer good overviews of the subject. Most of the major commercial publishers such as Marcel Dekker, McGraw Hill, and Prentice Hall publish monographs on CAD, CAM, CIM and computer graphics. The Society of Manufacturing Engineers and its Computer and Automation Systems Association are major contributors to the CAD/CAM monographic literature.

*Manufacturing Engineering and Materials Processing* and *Mechanical Engineering* are Marcel Dekker monographic series that include several volumes on CAD/CAM. The *SME Technical Paper Series* offer short papers on CAD/CAM. In addition to its *Technical*

*Paper Series*, SME publishes *Manufacturing Update Series* and *Productivity Equipment Series*. *IDC Computer Integrated Manufacturing Services Series* published by International Data Corporation is chiefly a marketing report series.

### *Societies and Research Centers*

The key organizations for CAD/CAM are the Society of Manufacturing Engineers, the Computer Automation and Systems Association of the SME, the IEEE Computer Society, the International Federation for Information Processing Computer Applications Technical Committee and Computer-Aided Manufacturing-International. In addition to sponsoring workshops and conferences, all support vigorous publication programs on CAD/CAM. SME also

FIGURE 5

Citing Journal Titles

| Frequency of Citing | Impact Factor | Journal Title |
|---|---|---|
| 15 | 1.211 | IEEE Transactions on Microwave Theory and Techniques |
| 10 | 2.342 | IEEE Transactions on Electron Devices |
| 9 | 1.033 | IEEE Transactions on Circuits and Systems |
| 8 | - | IEEE Proceedings |
| 5 | 1.327 | Electronics Letters |
| 5 | 0.882 | IEE Proceedings D |
| 5 | 0.462 | IEE Proceedings G |
| 4 | 1.668 | IEEE Transactions on Computer-Aided Design of Integrated Circuits and Systems |
| 4 | 0.847 | IEEE Transactions on Sonics and Ultrasonics |
| 3 | 1.004 | IEEE Journal of Solid State Circuits |
| 3 | 0.717 | International Journal of Control |
| 3 | - | Philips Technical Review |

offers a technical information retrieval service, CARIC. The *Encyclopedia of Associations* and *Directory of Engineering Societies and Related Organizations* list professional organizations.

There are over fifty research centers for CAD or CAM in the United States. Most are affiliated with universities. Cornell University's Manufacturing Engineering and Productivity Center, MIT's Laboratory for Manufacturing, and RPI's Center for Manufacturing Productivity and Technology Transfer are some examples.

The National Bureau of Standard's National Engineering Laboratory has a number of facilities involved in CAD/CAM research. These include the Automated Manufacturing Research Facility and the Center for Manufacturing Engineering which has several subdivisions. The Air Force Integrated Computer-Aided Manufacturing Program (ICAM) and NASA's Integrated Program for Aerospace-Vehicle Development (IPAD) are other government programs devoted to CAD/CAM research. The *Directory of Government Research* and the *Scientific and Technical Organizations and Agencies Directory* list government research programs. In addition to the Gale directories, both the *Computer Graphics Marketplace* and the *Robotics, Computer-Aided Design Market Place* include listings for major societies and research programs in the field.

### Standards and Specifications

The *Computer Graphics Handbook* provides an excellent explanation of various adopted and proposed computer graphics standards. Although limited to six standards, the handbook gives a history of each standard and information on how to obtain copies of standards. CAD/CAM and computer graphics standards can best be identified using online databases such as BRS's *Voluntary Standard Information Network*, *Military & Federal Specifications & Standards*, or *Industry and International Standards* and DIALOG's *Standards & Specifications*.

### Patents

United States patents can be identified using BRS's *PATDATA*, DIALOG's or SDC's *CLAIMS/U.S. PATENTS*, or Pergamon's *PATSEARCH*. *CAD/CAM Abstracts*, a monthly current awareness tool, lists new patents. Other sources for patents are *The Patent Newsletters*, *Catalog of Government Patents*, and *NASA Patent Abstracts Bibliography*. Since 1984 *Mechanism and Machine The-*

*ory* has included abstracts from Pergamon's *PATSEARCH* database. *Mechanism and Machine Theory* also offers document delivery for patents it lists. Most of the online services also offer document delivery services. Coverage of foreign patents is provided by Japan Patent Information Organization and DIALOG's or SDC's *WPI (World Patent Index)*.

### *Software*

Software directories abound. *International Software Directory* (Imprint Software) and *Software Catalog* (Elsevier) are multidisciplinary directories listing CAD/CAM software. The most timely sources for CAD/CAM software are online databases such as DIALOG's *Menu: The International Software Database* or BRS's *Online Microcomputer Software Guide and Directory*. Both list programs for personal computers. Engineering Software Exchange publishes *Computer-Aided Design Systems Update* which lists software. Tab Books publishes *International CAD/CAM Software Directory*.

A number of professional organizations also offer software. Most trade publications include software columns and a number of research journals have begun to do the same. The trade journals also include advertisements for CAD/CAM systems which can be useful in identifying software.

## NOTES

1. Analog computers aid plane design. *Aviation Week*. 53(8):37; 1950 August 21.
2. Ralston, Anthony; Reilly, Edwin D., Jr. *Encyclopedia of computer science and engineering*. 2d ed. New York: Van Nostrand Reinhold; 1983: p.1231.
3. New computer language makes 3-axis contouring a reality. *Machinery*. 67(12):139; 1961 August.
4. Smith, Christopher. The computer in design engineering. *Mechanical Engineering*. 86(4):34; 1964 April.
5. Computer-developed designs drawn electronically. *American Machinist/Metalworking Manufacturing*. 105(4):102; 1961 February 20.
6. Smith. Computer in design engineering: p.31.
7. Prince, M. David. Man-computer graphics for computer-aided design. *Proceedings of the IEEE*. 54(12):1701; 1966 December.
8. Coons, Steven Anson. An outline of the requirements for a computer-aided design system. In: *1963 Spring Joint Conference*, AFIPS Conference Proceedings, vol. 23. Baltimore, MD: Spartan Books, Inc., 1963: p.299.
9. Smith, Christopher. Computer in design engineering; p. 29.
10. Maguire, Thomas. The computer: design assistant. *Electronics*. 36(20):16; 1963 May 17.

11. Prince. Man-computer graphics; p.1698.

12. Ibid.

13. Holstein, David. Are you ready for computer-aided design? *Product Engineering*. 35(24):73; 1964 November 23.

14. Ibid.

15. Raney, J. Philip; Weidman, Deene J. NASTRAN Overview: dynamics application, maintenance, acceptance. *Shock and Vibration Bulletin*. 42(5):109; 1972 January.

16. Krouse. *What every engineer should know about computer-aided design and computer-aided manufacturing: The CAD/CAM revolution*. New York: Marcel Dekker; 1982: p.12

17. Myers, Ware. Interactive computer graphics: poised for takeoff? *Computer*. 11(7):61; 1978 January.

18. Van Deusen, Edmund. *Graphics standards handbook*. Laguna Beach, CA: CC Exchange; 1985; p.1/5-1/6.

19. Staayer, David H. Adopting applictions to the graphical kernel system. *Computer Design*. 22(7):169; 1983 July.

20. Ibid., p.168.

21. Ibid., p.169.

22. Williams, Tom. Degree of PHIGS/GKS compatibility still hotly debated. *Computer Design*. 24(17):38; 1985 December.

23. Ibid., p.38.

24. Van Deusen. *Graphics Standards Handbook*. p.1/7.

25. Jones, Keith. International group prepares worldwide CAD/CAM standards. *Mini-Micro Systems*. 19(1):44; 1986 January.

26. Parker, Richard; Shapiro, Sydney F. Workstations that chip design from end to end. *Computer Design*. 22(7): 160; 1983 July.

27. Langhorst, Fred E. Working toward standards in graphics. *Computer Design*. 21(7):182; 1982 July.

28. Zajdel, Theodore T. Standardizing computer integrated manufacturing. *ASTM Standardization News*. 12(10):20; 1984 October.

29. English, C. H. Computer Aided Design-Drafting (CADD)—engineering/manufacturing tool. *Journal of aircraft*. 10(12):748; 1973 December.

30. Integral CAD/CAM system reaches off-the-shelf status. *Machine Design*. 46(24):10; 1974 October 3.

31. Ibid.

32. Post, Charles T. The strong link that ties CAD to CAM. *Iron Age*. 220(8):30; 1977 August 22.

33. Krouse. *What every engineer should know*; p.16.

34. Computer-aided design and manufacturing. *McGraw Hill encyclopedia of science and technology*. 5th ed. New York: McGraw Hill; 1982; v. 3, p.491-492.

35. Ibid.

36. Dorfman, Julius. CAD/CAM: past, present, and future. *Design News*. 41(5):125; 1985 March 4.

37. Schaffer, George. CAD/CAM revolution. *American Machinist*. 116(5):72; 1972 March 6.

38. CAM: ready, willing and able. *Manufacturing Engineering and Management*. 71(2):17; 1973 August.

39. Schaffer. CAD/CAM revolution; p.72.

40. CAD/CAM—the 3M experience. *Manufacturing Engineering*. 81(1):40; 1978 July.

41. Schaffer. Computers in manufacturing. *American Machinist*. 122(4):sr/5; 1978 April.

42. Krouse. *What every engineer should know*; p.5.

43. Byles, Torrey. CADCON West looks at the future of CAD/CAM/CAE. *IEEE Computer Graphics and Applications*. 5(3):80; 1985 March.

44. Pluhar, Kenneth. Discrete graphics systems builders fighting for leadership in an expanding market. *Control Engineering*. 32(6):55; 1983 June.

45. Krouse. *What every engineer should know*; p.20.

46. The interface challenge. *American Machinist*. 129(1):96; 1985 January.

47. Krouse. *What every engineer should know*. p.35.

48. Bell, D. A. Employment skills for the robot age. *Robotica*. 3(2):93; 1985 April-June.

49. Krouse. *What every engineer should know*; p.16-17.

50. Schaffer. Computers in manufacturing. p.115.

51. Gross national product by industry: 1970 to 1983. Table no. 718. In: *Statistical Abstracts of the United States*: *1985*. Washington, DC: United States Bureau of the Census; 1984; p. 433.

52. Cleavelend, Peter. CAD/CAM: a technology update: *Instruments & Control Systems*. 56(3):20; 1983 March.

53. Annual CAD/CAM industry. *Tooling and production*. 51(12):22; 1985 May.

54. CAE/CAD/CAM revenues top $5 billion in 1984. *IEEE Computer Graphics & Applications*. 5(5):81; 1985 May.

55. Christman, Allan M. Update on CAD, CAM and CIM. *I&CS—The Industrial Process and Control Magazine*. 57(5):54; 1983 May.

56. Krouse. *What every engineer should know*; p.51.

57. Ibid., p.135.

58. Ibid.

59. Ibid.

60. Interface challenge; p.99.

61. Zajdel. Standardizing computer integrated manufacturing; p.18.

62. Interface challenge; p.97.

63. Christman, Alan M. Update on CAD, CAM, and CIM; p.53.

64. Schaeffer, Harry G. Integrating the islands of automation. *IEEE Computer Graphics and Applications*. 5(2):16; 1985 February.

65. Users and vendors evaluate the current status and future prospects of CAD/CAM. *IEEE Computer Graphics and Applications*. 4(2):28; 1984 February.

66. Interface challenge; p.99-100.

67. Krouse. Automation revolutionizes mechanical design. *High Technology*. 4(3):38; 1984 March.

68. Christman. Update on CAD, CAM, and CIM; p.55.

69. Users and vendors evaluate the current status and future prospects of CAD/CAM. *IEEE Computer Graphics and Applications*. 4(2):18-29; 1984 February.

70. Cleaveland. CAD/CAM: a technology update; p.23.

71. Interface challenge; p.99.

72. Richards, Larry G. Engineering Education: a status report on the CAD/CAM revolution. *IEEE Computer Graphics and Applications*, 5(2):20; 1985 February.

73. Klomp, Chris W. Recent advances in CAD/CAM and future trends. *Hydraulics & Pneumatics*. 36(10):52; 1983 October.

74. Dixon, J. R. Computers that design: expert systems for mechanical engineers. *Computers in Mechanical Engineering*. 3(3):10; 1983 November.

75. Hayes-Roth, Frederick; Waterman, Donald A.; Lenat, Douglas B. *Building expert systems*. Reading, MA: Addison Wesley Pub.; 1980; p.401.

76. Interface challenge; p.95.

77. Design Automation Conference, 22nd, 1985, Las Vegas, NV. *Proceedings*. New York: Association for Computing Machinery and IEEE Computer Society; 1985; p.iv.

78. Journal ranking. *SCI Journal Citation Reports*. 15(1);3-38; 1984.

# BIBLIOGRAPHY

## Books

Design Automation Conference, 22nd, 1985, Las Vegas, NV. *Proceedings*. New York: Association for Computing Machinery and IEEE Computer Society; 1985.

Krouse, John K. *What every engineer should know about computer-aided design and computer-aided manufacturing: The CAD/CAM revolution*. New York: Marcel Dekker; 1982.

Hayes-Roth, Frederick; Waterman, Donald A.; Lenant, Douglas B. *Building Expert Systems*. Reading, MA: Addison Wesley Pub. Co.; 1980.

Hubbard, Stuart W. *The computer graphics glossary*. Phoenix, AZ: Oryx Press; 1983.

Ralston, Anthony; Reilly, Edwin D., Jr. *Encyclopedia of computer science and engineering*. 2d ed. New York: Van Nostrand Reinhold; 1983.

*Statistical Abstracts of the United States: 1985*. 105th ed. Washington, DC: United States Bureau of the Census; 1984.

Van Deusen, Edmund. *Graphics standards handbook*. Laguna Beach, CA: CC Exchange; 1985.

## Articles

Analog computers aid plane design. *Aviation Week*. 53(8):37-38; 1950 August 21.

Annual CAD/CAM industry. *Tooling and Production*. 51(12):22, 26, 30; 1985 May.

Bell, D. A. Employment skills for the robot age. *Robotica*. 3(2):93-95; 1985 April-June.

Borda, Daniel J. CID: new tool for new times. *Consulting Engineer*. 60(40):70-73; 1983 April.

Breedon, D. B. Comparing analog and digital computers for solving design problems. *Machine Design*. 28(25):140-144; 1956 December 13.

Burgam, Patrick M. Extending CAD/CAM to a manufacturing community. *Manufacturing Engineering*. 92(3):40; 1984 March.

Byles, Torrey. CADCON West look at the future of CAD/CAM/CAE. *IEEE Computer Graphics and Applications*. 5(3):80-83; 1985 March.

CAD/CAM—the 3M experience. *Manufacturing Engineering*. 81(1):40-41; 1978 July.

CAE/CAD/CAM revenues top $5 Billion in 1984. *IEEE Computer Graphics & Applications*. 5(5):81; 1985 May.

CAM: ready, willing and able. *Manufacturing Engineering and Management*. 71(2):17; 1973 August.

Christman, Allan M. Integration of CAD and CAM: no longer a house divided. *Design News*. 40(21):109-114; 1984 November 5.

Christman, Allan M. Update on CAD, CAM and CIM. *I&CS—The Industrial Process and Control Magazine*. 57(5):53-57; 1984 May.

Cleaveland, Peter. CAD/CAM: a technology update. *Instruments & Control Systems*. 56(3):20-24; 1983 March.

Computer-aided design and manufacturing. *McGraw Hill Encyclopedia of Science and Technology*. 5th ed. New York: McGraw Hill, 1982; v.3; p.490-492.

Computer-developed designs drawn electronically. *American Machinist/Metalworking Manufacturing*. 105(4):102; 1961 February 20.

Computer graphics key to CAD/CAM. *American Machinist*. 123(9):77; 1979 September.

Coons, Steven Anson. An outline of the requirements for a computer-aided design system. *1963 Spring Joint Conference*, AFIPS Conference, *Proceedings*. vol. 23. Baltimore, MD: Spartan Books, Inc.; 1963; p.299-304.

Dixon, J. R. Computers that design: the expert systems for mechanical engineers. *Computer in Mechanical Engineering*. 3(3):10-18; 1983 November.

Dorfman, Julius. CAD/CAM: past, present and future. *Design News*. Part 1 in 41(4):171-178; 1985 February 18 and Part 2 in 41(5):125-130; 1985 March 4.

English, C. H. Computer aided design-drafting (CADD)—Engineering/Manufacturing Tool. *Journal of Aircraft*. 10(12):747-752; 1973 December.

For drafting or digitizing. *Automation*. 12(9):10, 12; 1965 September.

Graphic display in computers becomes more useful in design. *Product Engineering*. 37(20):51-52; 1966 September 26.

Hegland, Donald E. The many faces of CAD/CAM. *Production Engineering*. 26(6): 56-69; 1979 June.

Holstein, David. Are you ready for computer-aided design. *Product Engineering*. 35(24):66-76; 1964 November 23.

Integral CAD/CAM system reaches off-the-shelf status. *Machine Design*. 46(24):10; 1974 October 3.

Intel unveils iSBC 569 intelligent digital controller. *Manufacturing Engineering*. 84(1):61-78; 1980 January.

The interface challenge. *American Machinist*. 129(1): 95-102; 1985 January.

Is CAD worthwhile? *Machine Design*. 55(19):90; 1983 August 25.

Jadrnicek, Rik. Computer-aided design. *Byte*. 9(1):172-209; 1984 January.

Jones, Keith. International group prepares worldwide CAD/CAM standards. *Mini-Micro Systems*. 19(1):44; 1986 January.

Klomp, Chris W. Recent advances in CAD/CAM and future trends. *Hydraulics & Pneumatics*. 36(10):52; 1983 October.

Krouse, John K. Automation revolutionizes mechanical design. *High Technology*. 4(3):36-45; 1984 March.

Krouse, John K. Training CAD/CAM operators. *Machine Design*. 55(21):79-81; 1983 September 22.

Kunii, Tosiyasu. Practice and progress in CAD/CAM. *Computer*. 17(2):11; 1984 December.

Learning CAD/CAM. *Mechanical Engineering*. 105(5):45; 1983 May.

Langhorst, Fred E. Working toward standards in graphics. *Computer Design*. 21(7):177-178, 180, 182; 1982 July.

Maguire, Thomas. The computer: design assistant. *Electronics*. 36(21): 16-17; 1963 May 17.

Myers, Ware. Interactive computer graphics: poised for takeoff? *Computer*. 11(7):60-78; 1978 January.

New computer language makes 3-axis contouring a reality. *Machinery*. 67(12):138-139; 1961 August.

Parker, Richard; Shapiro, Sydney F. Workstations that chip design from end to end. *Computer Design*. 22(7):143-160; 1983 July.

Pluhar, Kenneth, Discrete graphics systems builders fighting for leadership in an expanding market. *Control Engineering*. 30(6):55-59; 1983 June.

Post, Charles T. The strong link that ties CAD to CAM. *Iron Age*. 220(8):29-31; 1977 August 22.

Prince, M. David. Man-computer graphics for computer-aided design. *Proceedings of the IEEE*. 54(12):1698-1708; 1966 December.

Raney, J. Philip; Weidman, Deene J. NASTRAN Overview: dynamics application, maintenance, acceptance. *Shock and Vibration Bulletin*. 42(5):109-127; 1972 January.

Richards, Larry G. Engineering education: a status report on the CAD/CAM revolution. *IEEE Computer Graphics and Applications*. 5(2):19-25; 1985 February.

Ross, Douglas T. The APT joint effort. *Mechanical Engineering*. 81(5):70; 1959 May.

Schaeffer, Harry G. Integrating the islands of automation. *IEEE Computer Graphics and Applications*. 5(2):16-17; 1985 February.

Schaffer, George. CAD/CAM revolution. *American Machinist*. 116(5):72; 1972 March 6.

Schaffer, George. Computers in manufacturing. *American Machinist*. 122(4):sr/1-sr/5; 1978 April.

Siders, Ronald A. Computer-aided design. *IEEE Spectrum*. 4(11):84-92; 1967 November.

Simon, Richard L. The marriage between CAD/CAM systems and robotics. *Design News*. 39(21):87-93; 1983 November 7.

Smith, Bradford M. IGES: a key to CAD/CAM systems integration. *IEEE Computer Graphics and Applications*. 3(1):78-83; 1983 November.

Smith, Christopher. The computer in design engineering. *Mechanical Engineering*. 86(4): 29-35; 1964 April.

Staayer, David H. Adopting applications to the graphical kernel system. *Computer Design*. 22(7):167-172; 1983 July.

Stotz, Robert. Man-machine console facilities for computer-aided design. *1963 Spring Joint Conference. AFIPS Conference Proceedings*, vol. 23. Baltimore, MD: Spartan Books; 1963; p.323-328.

Users and vendors evaluate the current status and future prospects of CAD/CAM. *IEEE Computer Graphics and Applications*. 4(2):18-29; 1984 February.

Williams, Tom. Degree of PHIGS/GKS compatibility still hotly debated. *Computer Design*. 24(17):38-39; 1985 December.

Zajdel, Theodore T. Standardizing computer integrated manufacturing. *ASTM Standardization News*. 12(10):18-21; 1984 October.